FIELD GUIDE TO THE PATCHY ANTHROPOCENE

FIELD GUIDE *to the* PATCHY ANTHROPOCENE

THE NEW NATURE

Anna Lowenhaupt Tsing
Jennifer Deger
Alder Keleman Saxena
and
Feifei Zhou

STANFORD UNIVERSITY PRESS
Stanford, California

Stanford University Press
Stanford, California

Printed in the United States of America on acid-free, archival-quality paper

Library of Congress Cataloging-in-Publication Data
Names: Tsing, Anna Lowenhaupt, author. | Deger, Jennifer, author. | Saxena, Alder Keleman, author. | Zhou, Feifei, author.
Title: Field guide to the patchy Anthropocene / Anna Lowenhaupt Tsing, Jennifer Deger, Alder Keleman Saxena, and Feifei Zhou.
Description: Stanford, California : Stanford University Press, 2024. | Includes bibliographical references and index.
Identifiers: LCCN 2023028438 (print) | LCCN 2023028439 (ebook) | ISBN 9781503637320 (cloth) | ISBN 9781503638662 (ebook)
Subjects: LCSH: Nature—Effect of human beings on. | Global environmental change. | Human ecology. | Environmental degradation. | Geology, Stratigraphic—Anthropocene.
Classification: LCC GF75 .T75 2024 (print) | LCC GF75 (ebook) | DDC 304.2—dc23/eng/20231016
LC record available at https://lccn.loc.gov/2023028438
LC ebook record available at https://lccn.loc.gov/2023028439

Cover design: Michele Wetherbee
Cover art: Feifei Zhou, with Amy Lien and Enzo Camacho, *Acceleration*. Image and drawings from *Feral Atlas: The More-Than-Human Anthropocene*.

CONTENTS

PART IV

FIELD GUIDE
TO THE
PATCHY
ANTHROPOCENE

Introduction

The future arrives almost every day now
each time with dizzying force

where—and how
—are we to find our bearings?

Human action has transformed the planet—from its climate, to its life forms, to its hydrology and chemistry. Such changes have led to the use of the term "Anthropocene" as an epoch of human-influenced environmental change. The Anthropocene is planetary. Yet, as this book will show, it is also up close and personal. It is impossible to understand the Anthropocene without finding out what is happening in particular places: the patches of our title. This is not because planetwide conditions have local *effects* and *instantiations*, as so many studies have it; instead, as this book argues, patches *make* the Anthropocene, including its planetary forms. While big data modeling has usefully brought the idea of the Anthropocene into play, learning more about it requires descriptive field studies and histories, that is, fine-grained attention to patches. What observers find, too, may continue to surprise us: At every scale, nature is changing through its entanglements with human infrastructure. The new Nature is feral. How shall we as observers re-attune ourselves? A field guide can help.

Consider the red turpentine beetle, *Dendroctonus valens*, as described by biologist Scott Gilbert.[1] This beetle lives under the bark of damaged pine trees, drilling out tunnels and creating brood galleries, which can kill the tree by destroying its vascular system. In its native North America, this beetle doesn't do much harm, even when it is covered with the fungus *Leptographium procerum*, which helps the beetle drill by digesting the wood. The beetle only came to global attention when it hitchhiked on industrial shipments of unprocessed timber from North America to China's Shanxi province in the 1980s. There, the beetle was able to form an association with a much more potent local variant of its symbiotic fungus, *L. procerum*. Working together, this potent local fungal variant and this imported beetle started killing trees in droves. They even roped the trees into bringing on their own demise; the fungus stimulated the trees to produce volatile chemicals that attracted more fungus-covered death-dealing beetles. Gilbert reports that more than 10 million trees have been killed in China, stalling reforestation efforts. Looking ahead, it seems likely that the potent Shanxi fungus will coat beetles that make the return trip, again on industrial timber, to North America. Gilbert addresses U.S. readers as he quotes a pest management official: "Be afraid."

The red turpentine beetle carries the story of this book in several ways. First, its appearance in China depends entirely on the industrial trade in timber; the beetle would have no way to make it from North America to China without this exchange. In the vocabulary of this book, the Chinese occurrence of the beetle is *feral*—that is, transformed by human infrastructure but not under the control of human designers. The term "infrastructure" here refers to landscape-changing human building projects, such as the ships, roads, and warehouses that move timber around the world. The beetles become feral as they take advantage of the global timber trade.

The concept of ferality (to which we return later in this Introduction) requires thinkers to pay attention to both human arrangements and nonhuman response. Calling the beetle feral signals that, instead of dismissing trade as a necessary part of being human, we need to study the particular conditions and routes of the commodity chain. Calling the beetle feral also focuses attention on the beetle itself, as it has attracted a new fungal ally in its Shanxi tunnels. Just how do beings transform in their responses to human infrastructures, and how do they remake the world? The term "feral" forces attention to what was once hidden in the gap between "wild" and "domestic."

The red turpentine beetle carries a second point of the book. Together with their fungal allies, the beetles help make the Anthropocene *precisely* because of their patchy occurrence. The beetle comes from North America. The potent fungal variety lives in Shanxi, China. These places matter to the story: It is the transfer of the beetle to a new place that sparks its alliance with a potent local fungus, in turn leading to the death of millions of trees. The phenomenon of tree death in this story comes into being because of the patchy distribution of beetles and fungi. It is this too that sparks the fears of North Americans, who expect the beetle to come back with deadly new powers. Patches and corridors are central to the remaking of the earth that characterizes the Anthropocene. This book shows how to bring a spatial analysis—that is, an analysis of patches and corridors—into description of the Anthropocene.

The heterogeneity of time is also important. The *history* of the beetle, as it encounters the potent fungus, matters; this encounter happened not only in a particular place but also at a particular time—for the transnational timber trade, for Shanxi reforestation efforts, and for fungal evolution and dispersion. Meanwhile, this history unfolds in the shadow of the possibility of an ecosystem rupture, should the forests in this region be deprived of a whole species, along with the insects, fungi, mosses, and other beings that depend on that species.[2] Temporal coordinations—and discontinuities—are key elements of the Anthropocene.

Complexes of human infrastructure, we argue, rupture the temporal coordinations that sustain multispecies life. Infrastructures produce feral effects because they modify land, water, and air, causing responses from all the beings that encounter these modifications. Throughout the history of our species, humans have been engaged in landscape modification projects—such as burning to create grazing spaces for game, making small dams for fishing, encouraging some plants over others, and much more. Precolonial kingdoms encouraged much more radical infrastructure projects, such as remaking rivers, but these tended to be limited to areas under direct regional governance. To consider how humans have become a geological force—moving more dirt than the Ice Age glaciers, causing faster extinctions than anything but a meteor, and changing the climate in a way that may take millions of years to readjust—something more than just "being human" comes into play. The imperial expansion of Europe, on the one hand, and capitalist investment, on the other, radically changed the impact of the human presence on earth through massive infrastructure-building projects, which modified land, water, and atmosphere in an unprecedented manner. Feral effects follow such radical transformations. This book argues that imperial and industrial infrastructures of the last five hundred years should be singled out for their exceptional force in changing the earth's surface.[3] The power of their political economies has allowed terraforming at a scale that surpasses previous human interventions. This terraforming is dangerous to life on earth, human and nonhuman. It is for this reason that we imagine this book as a set of materials for "unbuilding." Dismantling huge swaths of imperial and industrial infrastructure might be our best chance for sustaining more-than-human livability.

One more point of this book rides on the red turpentine beetle: Attention to the beings, places, ecologies, and histories of the Anthropocene can reignite curiosity and wonder concerning our damaged planet, even as we denizens may strive to make things different. That's why we have produced *Field Guide*. We want to inspire readers to look around in new ways. Gilbert explains how the red turpentine beetle and the fungus together form a monstrous conception only newly recognized: a "holobiont," an assemblage of beings that evolves and adapts together. Earlier traditions of biology and natural history glanced over such beings, with their multispecies complexity. This and other forms of excitement await readers who dare to explore further.[4]

Why a Field Guide?

Field guides teach us how to notice, identify, name, and so better appreciate more-than-human worlds. They provide the descriptive details by which to hone our powers of observation, while offering access to specialized information that we may not be able to glean on our own. To head out into the world with a field guide in hand is to commit to opening and attuning in new ways. It is the start of a relational practice of coming to knowledge that holds the potential for cultivating deeper connections with the rhythms, relationships, and vulnerabilities of more-than-human worlds: an art of noticing that requires you to develop comparative capacities, together with an appreciation of form, color, texture, movement; a mode that marks a deliberate commitment to reorient yourself in the world; an exercise that can lead to you noting things being both in and out of place; a habit attuned to the rhythms of flourishing and diminishment; an attention that enables you to notice and miss critters no longer present, along with what remains and accrues in their stead.

We made *Field Guide* with the conviction that, if we are to learn to better reckon with these times of human-caused environmental catastrophe, it is essential to reflect more closely on *how we got to where we are.* There is, of course, no one answer to this question. Likewise, there is no singular "we" or "where" with which to adequately account for what is going on—not to mention *what is at stake and for whom and in what ways*—as ecologies lurch and stumble across the planet.

This book shows how both spatial and temporal heterogeneity can be useful when analyzing the Anthropocene. In the practices we advocate, on-the-ground social ecologies are at least as important as the thought experiments, such as general circulation climate models, used to imagine a planetary scale. This involves more than taking conventional forms of scholarship and slotting them into expected roles. Rather, as we will show, the analysis of a "patchy Anthropocene" makes possible a new form of attention to the particular in the planetary—and the planetary in the particular.[5]

Decolonizing Natural History for the Anthropocene

Have a look at the cycad palm. . . . That's my *märi* [mother's mother or mother's mother's brother]. It belongs to the land, stories and songs of my mother's mother clan, the Wangurri people. *Märi* is the boss for ceremony and everything, really. *Märi* is our backbone. That's how it is for us. That's gurruṯu, what you call family or kinship.

—*Yolŋu elder Paul Gurrumuruwuy*

Over recent decades natural history has fallen out of favor in the sciences, where the rise of big data has substantively recast the grounds of research and analysis. Satellite views and composite datasets can seem to render on-the-ground observations quaintly out of scale in relation to planetary matters. Recently, however, and no doubt in response to public awareness of environmental damage, natural history has been undergoing an astonishing resurgence.

FIGURE 1. **Paul Gurrumuruwuy's *märi*.**
MIYARRKA MEDIA

As biology professor and citizen of the Potawatomi Nation Robin Wall Kimmerer suggests, to address the increasingly urgent question of *How are we to live?* we must start paying more attention to the world around us.[6] As we write this book, new kinds of natural history writing abound—films, websites, and exhibitions also.[7] Across the humanities, sciences, and arts, new kinds of transdisciplinary research and advocacy projects find energy and purpose in the work of cultivating attunements to situated ecological relations, fostering sometimes radical reorientations towards more-than-human worlds of vitality and coexistence.[8] Indeed, as the social sciences and humanities continue to come to terms with natural history's central role in colonial and imperial projects, natural history museums have become important sites for interventions that directly challenge modernist knowledge systems and the values that they instantiate. Not all of these interventions have the same commitment to decolonizing knowledge that we offer here. However, we believe we can work in dialogue with these projects, expanding them to offer alternative, plural natural histories.[9]

Natural history offers arts of noticing that can draw practitioners beyond blinding narratives of progress to attend to the temporalities of those living and nonliving others who share the planet with humans—and who make it possible for humans to survive. The natural history that we advocate here recognizes nonhumans—such as red turpentine beetles—as robust historical protagonists in their own right. In doing so, it contributes to an urgent recalibration of the categories of both nature and history, not to mention the long-overlooked relationship between the two. Bringing forth perspectives grounded in dynamically interacting material histories, the Anthropocene natural history we promote is always social and cultural. Indeed, we take this as a fundamental lesson of the Anthropocene. In promoting a revitalized natural history, we do not skirt the fact that the history of natural history remains indelibly implicated in the brutal avarice and racist typologies of colonialism. One means of acknowledging this heritage has been to make the work of identifying ongoing histories of invasion, dispossession, and extraction central to our methods.

We also introduce other modes for decolonizing natural history. First, we reach out to a variety of legacies and practices of observation, rather than limiting observation to Western scientists. We acknowledge non-Western observers of many kinds, as well as many kinds of Indigenous and vernacular admirers of the natural world.[10] Our natural history is wonderfully contaminated with a diversity of forms of empiricism. It thus offers a critical eye on imperial traditions that imagine themselves universals. Second, our natural history is always combined with social history—including issues of social and environmental justice.[11] We do not aim to segregate the natural from the cultural; instead, we explore how Anthropocene patches are created through allowing social injustice to shape environmental catastrophe. Third, rather than allowing the arts—including the art of mapping—to merely illustrate our social-natural history, we call on them to define and conceptualize problems. By refusing the conventional aesthetics of an unrooted, context-free art, on the one hand, and expanding the sensual grip of the sciences, on the other, we show forms through which situated artists and creative scholars might co-create knowledge. Map-making in particular is a method for learning about the Anthropocene.

Our method to work across modes of doing natural history is to "pile" many forms together. Instead of working toward a single common language, we juxtapose and amass multiple kinds of empiricism. "Piling" is a knowledge practice that embraces heterogeneous forms of Anthropocene knowledge, assembled without the imposition of preordained theory or established disciplinary hierarchies. While this is not the only way to decolonize natural history (see Part IV), it is an opening to the many natural histories that surround us, if we only allow ourselves to notice them.

In the process of looking at patches, we hope users of *Field Guide* might find "Holocene fragments," to use Zachary Caple's term for patches in which longtime evolutionary mutualisms, such as arrangements that link pollinators and flowering plants, remain healthy.[12] Natural history of the sort we advocate can find thriving patches as well as terrifying ones, and it is important to be able to tell the difference. In this way, too, we might begin to think about limiting the spread of the worst Anthropocene patches, while preserving and nurturing refuges for Holocene ecosystems. Learning to observe and describe the patchy Anthropocene allows us to notice processes, protagonists, and connections that could show us how to stop Anthropocene proliferations.

We use the term "field guide" in this book, then, in a rather literal sense, not just as a metaphor. Our goal is to allow readers to understand the Anthropocene through the particularity of observations at multiple scales, while also inspiring readers to indulge in this practice themselves, observing the world through its Anthropocene patches. Even when we turn to larger, planet-spanning movements, our analysis stays close to particular organisms, ecological assemblages, and landscapes. It also shows how established conventions of natural history—as inherited from Europeans caught up in colonial adventures—are not enough.

Storytelling and Noticing: A Feminist Approach to the Anthropocene

Ursula Le Guin's essay "The Carrier Bag Theory of Fiction" offers important guidance for the kinds of stories needed for a field guide to the patchy Anthropocene.[13] Le Guin contrasts stories of Man the Hunter, where the action proceeds on the point of a spear, with those that offer the rhythms of women's gathering activities, assembling diverse stories. The former offer singular big histories; the latter pay attention to heterogeneity and contingent effects. Our *Field Guide* offers a carrier bag of stories, as is appropriate for a guide to the patchy Anthropocene. This does not mean that we abandon the big picture; on the contrary, we constantly search for structural and out-of-scale effects. But we see these produced through the dynamics of patches, rather than as the preformed effects of planetary algorithms.

In the social sciences, carrier bag approaches have been associated with theories of articulation, especially in feminist approaches to capitalism. As J. K. Gibson-Graham has argued, masculinist approaches were intent on producing master narratives of a singular all-powerful capitalism; they neglected the on-the-ground dynamics through which diverse economies have been created and linked.[14] The "Gens Manifesto" and its interlocutors followed up on this insight to show how articulations between kinship and industrial organization can shape the histories through which capitalist political economy develops.[15] In this spirit, instead of regarding planetary climate change, assessed through modeling, as the only index of Anthropocene transformations, we consider how human infrastructures in particular times and places remake species, geologies, and, more generally, environments. This leads to stories grounded in place and time: thus, a field guide.

A field guide approach is a way of doing "theory." Theory here provides traction with the material under analysis, rather than being imposed on it from without, as dogma. Free of the goal of theoretical purity, a variety of explanatory tools can be deployed to illuminate this book's stories. Imagine the scene of gathering called up by Le Guin. Digging an edible root or catching a frog each requires different tools. Theory enables our storytelling practice as tools to notice the patches, corridors, edge effects, and articulations of the Anthropocene. This means, too, that we tell stories through lots of details. The details matter in getting the story right.

Throughout *Field Guide*, readers will learn of particular places, times, and forms of life. Look to this for inspiration to find scenes other than those we identify here. Use it to gather your own stories.

More About the Feral

Field Guide is written in dialogue with our digital project, *Feral Atlas: The More-Than-Human Anthropocene*.[16] (To learn more about how to use that project, see the Appendix.) *Feral Atlas* is a mapping project in which field reports, based on empirical observation, describe Anthropocene patches. *Field Guide* depends on *Feral Atlas* for many of its case studies, and it opens up the conceptual terrain of *Feral Atlas* for further discussion. *Field Guide* thus continues our exploration of the feral.

Ferality, in our definition, is the state of nonhuman beings engaged with human projects, but not in the way the makers of those projects designed. Our definition draws on, but extends, the common use of the term "feral." A pig is considered feral when it leaves the farm to live in the forest. (Note that ferality is not necessarily a species characteristic: The pig on the farm has the same genetic signature as the feral pig.) We only need to push this term slightly to see the big-leafed vine kudzu as feral when it leaps beyond its planted spaces to smother houses and trees. But we also extend the term to nonliving entities. Carbon dioxide, for us, is feral when it moves out from fossil fuel–burning industries. Here, the feral entity, carbon dioxide, is identified by its non-designed anthropogenic effects: accumulating at atmospheric concentrations that can be neither controlled nor ignored.

Ferality is important for a field guide to the patchy Anthropocene because it shows the autonomous activities of nonhumans, even those most closely affected by human infrastructures. Without the concept of the feral, it is too easy to fall into a dichotomy that only includes the wild and the domestic. Wild things are imagined as having nothing to do with humans; domestics are imagined as entirely under human control. In the patchy Anthropocene, many nonhumans (living and nonliving) are responsive to human actions without submitting even slightly to human control. It is this set of beings that we call feral.

The concept of the feral is designed to urge readers to acknowledge the relationship between human infrastructures and the beings that respond to them: Both deserve careful study. Conventionally, natural scientists have studied beings without bothering with the anthropogenic relationships within which they are entangled. Conventionally, social scientists and humanists have ignored the activities of nonhumans except as they play into human plans. Feral beings require attention on both sides; paying attention to the feral forces readers into the entanglement of human and nonhuman trajectories.

As should be evident from our definition, designation as feral does not indicate whether we approve or disapprove of a particular bit of feral action. Trees that grow up in an abandoned lot are feral—and wonderful for many ecological reasons. Pathogens that evolve resistance to antibiotics are feral—and terrible for the humans likely to die of infections. "Feral" is not a term of political assessment. Our definition is thus quite different from alternative uses, such as the ferality glorified as unbridled masculinity, or condemned as unbridled chaos.[17]

In these times, noxious feral effects have become particularly evident, and we spend a lot of time with them in *Field Guide*—even while not ruling out the possibility of benign or beneficial feralities. In contrast to perspectives on the Anthropocene that imagine a time of human triumph, this book charts the dangers of human-induced environmental effects.[18] The Anthropocene is a time of great trouble, and it would be great if this epoch might be as short as possible. For the authors of *Field Guide*, recognizing and bearing witness to the ways that more-than-human assemblages generate environmental harm is a necessary first step to understanding the social and environmental disruptions that make up the Anthropocene.

An important caveat: We do not use the term "feral" to refer to humans or to human technologies and infrastructures. Vulnerable people have always been accused of unnecessary wildness, and we will not join that chorus. Human technologies and infrastructures, for us, are the apparatus that produce ferality.[19] Furthermore, whether or not they are adequate to meet the human needs they were designed for is not relevant to our analysis here. Ferality moves beyond human goals. Properly functioning and poorly functioning infrastructure (from a human standpoint) can each have feral effects.

Feral action is traced throughout *Field Guide*—in the identification of geographical patches, in the unexpected effects of human infrastructure, and in the shifting biologies of organisms reacting to the human presence. In each part of the book, we further develop the concept of the feral for its guidance in describing the Anthropocene.

Who Are We?

Any project about planetwide processes makes one think about the use of "we" pronouns. This book uses "we" in two different ways. Most of the time, when you read "we," think of the co-authors of this book. Occasionally, however, we, the authors, call up our readers as "we." The first-person plural can be problematically presumptive, hailing exclusive elites. At the same time, new forms of we are in the making—and the authors need them to address the epistemically patchy Anthropocene this book describes. Following Caribbean American literary critic Sylvia Wynter (see Chapter 1), this book aims to open up the we of the human condition.[20] Our readership we is utopian, reaching toward a not-yet-achieved diversity of patchy perspectives and overlapping concerns. Our we is an invitation.

Indeed, the we of the co-authors remains a collective in the making: a shared voice on paper composed by three anthropologists and an architect with divergent disciplinary orientations and experiences. In a nod to this, while most of the chapters that follow are co-authored, you will also find single-author contributions. In Chapter 5, Alder Keleman Saxena shows how unseasonal weather can reveal the inherent patchiness of climate change. In Chapters 7, 8, and 9, Anna Tsing dives deeper into the historical dynamics of the more-than-human Anthropocene, offering analytic tools for tracking crucial interactions between human and nonhuman histories that crisscross the humanities and sciences. In Chapter 11, Feifei Zhou argues for an attention to processes of both building and unbuilding in confronting the social and infrastructural damages of the Anthropocene; and in Chapter 12, Jennifer Deger offers a glimpse of co-creative practices with Yolŋu Aboriginal people to suggest an approach to collaborative knowledge that reaches beyond "piling." Writing this book together—sometimes integrating our voices in a single text, and sometimes putting our differing perspectives in conversation—has deepened our shared commitment to developing methodological and analytic tools for studying the emerging natural histories of the Anthropocene. Put another way, it has underscored our call for a "patchy epistemics," the subject of Part IV of the book.

For there is no way around it. While the world we are building was never ours alone, nonhuman histories now shape the future for us all. Learning to observe and describe the patchy Anthropocene in all its multi-sited, multi-scalar and more-than-human complexity is essential if we are to find ways to limit its world ripping force. Only through a sustained and collective effort of piling observations and insights—patch upon patch, perspective upon perspective—can we hope to identify, and so grapple with, the various protagonists, processes, and infrastructures through which the Anthropocene is being made with terrifying momentum. We don't just need a renewed natural history; we need many kinds of natural historians working in, and across, patches of all kinds to create new constellations of Anthropocene knowledge and collaboration.

Take *Field Guide* . . . and get out there.

FIGURE 2. **A stark landscape: abandoned overburden from brown coal mining in central Denmark, with a disintegrating cable in the foreground.**
ANNA TSING

PART I
Patches

This book asks observers to *treat the Anthropocene as patchy.* In this part, we argue for the importance of field observation and discuss our use of the patch, a concept from landscape ecology. Following patchy geographies shows us the Anthropocene up close.

Distinctive Anthropocene patches include human designed spaces such as suburbs and malls, as well as spaces stemming from human activities but not designed by humans, such as plastic gyres in the ocean. The location and extent of the gyres is created by the interaction of currents, winds, and plastic waste, rather than human design. Because there are no blueprints to study, observers need to describe and follow such patches. For example, to know the extent of the new diseases and toxic flows of the Anthropocene, observers must look around—rather than assuming the extent or seriousness of the problem in advance.

Chapter 1 lays foundations for learning to observe the patchy Anthropocene by putting our project in dialogue with other environmental humanists. Chapter 2 offers examples through which to identify Anthropocene patches. Chapter 3 opens up the idea of mapping so that Anthropocene patches of indeterminate contours might be tracked through mapping in an extended sense.

To get started on this journey, we offer here some photographs of a particular—and comparatively hopeful—Anthropocene patch: the sandy overburden left from brown coal mining in Søby Brunkulslejere, in central Denmark.[1] Because of the high water table and the unstable sands of the abandoned overburden, the area has not been developed. Much of the former mining area has now grown up in trees, both planted and self-seeded. The mining holes themselves have filled in, becoming acidic lakes. But there are still-remaining sand tips where overburden sands are exposed to the sun. Lodgepole pines, introduced in the 1970s, spread onto the empty sands through interactions with cosmopolitan mycorrhizal fungi, especially *Paxillus involutus* and *Pisolithus arrhizus*, which both do well in sites of human disturbance.

Anna Tsing participated in a project to study what grows from these sand tips; the images here are taken from that study.[2] The group was fascinated by the rush of life emerging in the abandoned overburden: An Anthropocene patch was on its way to regaining Holocene characteristics. Patches are rarely entirely without life for very long. But the kinds of life that flourish here are those particularly attuned to the affordances offered by human disturbance. In our brown coal mining site, disturbance-oriented species were creating a feral geography—and, indeed, a rather promising one from our perspective. These photographs introduce the strange and wonderful plants and fungi emerging from seemingly barren sands. Before turning to more dangerous patches, we offer a reminder of the benign possibilities of life gone feral.

The images tell of the pleasures of getting to know a patch, however disturbed, in all its intricacies.

FIGURE 3. **Even in the overburden, there is life. Lodgepole pines (*Pinus contorta*) self-seed in the sand.**

ANNA TSING

FIGURE 4. **Mushrooms emerge from symbiotic connections between pine roots and fungi; here *Paxillus involutus*.**

ANNA TSING

FIGURE 5. ***Pisolithus arrhizus*, "dead man's foot," is comfortable in artificial deserts.**
OLGA KOKCHAROVA

FIGURE 6. **The sand tips offer homes for plants of nutrient-poor places, such as *Drosera rotundifola*, a carnivorous sundew.**
ANNA TSING

FIGURE 7. **Our research team explored the connections between roots and fungi. Studying feral geographies involves getting one's hands dirty.**
ANNA TSING

FIGURE 8. **The feral geography of overburden includes both the sand and the life emerging from it.**
COLIN HOAG

FIGURE 9. **Red deer have gathered in this comparatively wild place. Here, the mushroom *Laccaria amethystina* grows next to a skull. Not all feral geographies are toxic; some are lively places.**
ANNA TSING

ONE

Bringing Field Observation to the Anthropocene

The term "Anthropocene" has been much criticized—and often rightly so—for what it omits as well as what it misleadingly emphasizes. Yet Anthropocene has become a gathering point for discussion of environmental dangers, across disciplines and political divisions, and this seems to us a good enough reason to use it.

Perhaps the one feature of the Anthropocene on which most commentators agree is its planetary nature: From climate change to the extinction crisis, and from radioactivity to industrial toxins, this is not a local storm. Such a planetary scale enables attention to region-crossing phenomena, such as climate. It is also a scale that allows for globe-crossing political mobilization. There is no place to hide from the Anthropocene, and this knowledge might yet prod policymakers to abandon the destructive pleasures of "business as usual." Yet many social justice advocates argue that the planetary scale, by itself, condemns the concept. It assumes a unitary humanity that simply doesn't exist. It posits equal responsibility by elites *and* the impoverished and oppressed. It deepens racial divisions—by ignoring them. Besides, given the high-tech modeling methods required to work at this scale, studying the planet as a whole privileges the work of "experts," who have earned public distrust in many quarters across the globe.

Both criticisms and claims for a planetary scale are worth taking seriously. What we argue is that just because the Anthropocene is planetary does not mean that it is *only* planetary.

Even the most dedicated planetary models are summations and projections from local phenomena. Models depend on the practice of "ground truthing," that is, testing projections against what is actually happening in particular places. This book turns the conventional relationship between models and ground truthing on its head: Taking ground truthing seriously creates a different kind of model, full of multiple, overlapping, empirical perspectives. By investigating the planetary effects of human world-making programs *at many different scales of space and time,* we show how it becomes possible to tell histories that reverberate across scales. Rather than disappearing into models, field observation comes to the fore; this is how we map landscape structures, which in turn help us think about systems, including planetary systems.

This is an insight from the environmental justice movement, which has shaped scholarship since the 1980s. In the United States, that scholarship started around the problems of toxic dumping in communities of color.[1] Studying this phenomenon required getting close to the dump itself as well as the community forced to live with it. But it also involved stepping back to see the larger racial ecology in which toxic wastes are differentially sited, and in which communities of color have been less able to get out of the way.[2] It is in moving across scales that environmental justice scholars combine the insights they gain from ordinary people's experiences and ecologies with that of a structural analysis of the landscape.[3] In the process of scale shifting, it becomes possible to consider how widespread Anthropocene phenomena, such as toxic dumps, are *constituted* in relation to racial inequalities. Rather than imagining a planet removed from social justice concerns, this kind of scale shifting allows the planet to come into focus as a product of our social history.

When the planetary nature of the Anthropocene is overemphasized, field-based observations disappear, banished from importance. Yet such banishment throws out the baby with the bathwater; we cannot possibly know the Anthropocene without describing it properly. Both humanists and natural scientists are part of the problem here. Too many natural scientists these days swoon over the pleasures of big data, which allow them to do research without leaving their offices, emphasizing planetary data sets over localized studies, and thus producing the kinds of scalable research most likely to secure funding. Humanists, for their part, often tend to denigrate description as old-fashioned and somehow, counterfactually, lacking the aristocratic distinction of "theory." In contrast to such dismissals, our team—in solidarity with similarly unconventional scientists and humanists—argues that *describing* the Anthropocene, in all its patches and processes, is the most exciting and challenging work of our times.

Field observations take many forms. The reports that our work draws on include both humans and nonhumans, involving history, ethnography, biology, geology, hydrology, and more. Artists also do important forms of field observation.[4] Within the academy, regional and subdisciplinary heritages of observation add to the variety of forms of empiricism. Furthermore, nonprofessionals have much to contribute. For instance, bird-watchers and mushroom collectors both draw from and contribute to the fields of ornithology and mycology. Others work from different knowledge traditions, such as Indigenous teachings and practices. Vernacular knowledge and experience are an asset, not a liability; many kinds of empirical observation are possible, each adding to knowledge. Indeed, it is in gathering across forms of observation that scholars, citizens, and advocates of many sorts might share curiosity and concern about the world. We have no chance of knowing the Anthropocene if we can't work together across these divides. Field-based curiosity can start conversation.

Our Interlocutor: An *Only* Planetary Anthropocene

Before showing how we use field observations to understand the Anthropocene, let's consider the alternative. Among environmental humanists, Dipesh Chakrabarty is perhaps the most devoted admirer of the planetary scale. Chakrabarty distinguishes himself from most other humanists by having carefully studied the relevant earth science and global modeling; he appreciates the *longue durée* narrative of the Anthropocene because he knows it. Chakrabarty was also one of the first humanists to value the Anthropocene as natural scientists presented it—rather than, as many humanists did in the first days of discussion, jumping immediately to criticism of the concept.[5] All Anthropocene scholars today are in his debt for this.

In explaining his appreciation of the planetary scale, Chakrabarty draws on his earlier work as a postcolonial scholar to show readers the philosophical possibilities of incommensurate knowledge systems. In this work, he differentiated between the Indian social history that most professionally trained historians presented, which he called History 1, and the temporalities of the Indian gods, which could not be accommodated in that history, and thus, for Chakrabarty, formed a separate History 2.[6] History 2 allows reflection on the omissions and distortions of History 1, even as it offers its own narratives.

In this same spirit, Chakrabarty asks humanists to separate the "global" and the "planetary" as two divergent kinds of history.[7] The global, a kind of History 1, tells the stories of politics and trade, and of subaltern status and rebellion. In contrast, the planetary, a kind of History 2, offers stories of the geological and climatic *longue durée*, stories that humble the human presence and refuse our petty distinctions. The Anthropocene, he argues, requires the latter stories; environmental humanists should not try to muddy the waters with their confusions between politics and climate. Instead, for Chakrabarty, the planetary can serve as reflexive commentary on more familiar humanist ways of knowing, such as the global.

This book follows Chakrabarty in appreciating the *longue durée* of geological time and the startling implications of taking the planet as a unit. However, in contrast, this book argues that it is possible to tell human histories that reverberate at many scales, including the geological and the planetary. Human histories *do* shape geological time. In spatial analysis, the planetary scale does not foreclose other scales; instead, it comes into being in the interplay across scales. In distinction from Chakrabarty's fears, this does not mean reducing the Anthropocene to a simple reflex of global politics. Instead, we urge attention to the patch ecologies through which both global politics and planetary geologies are created. Even global warming, the most planetary effect of the Anthropocene, emerges in the distinction between carbon dioxide-producing industrial centers and carbon dioxide-absorbing forest and ocean sinks. If we turn our attention to toxins, radioactivity, extinctions, and other Anthropocene phenomena, the patchy nature of exposure is even clearer. Justice issues, and thus politics, are clearly involved in the distribution of each of these threats. It is impossible to describe the Anthropocene without regard for the fundamentally unequal relations among humans through which the environmental effects of the human presence are structured.

More-Than-Human Field Observation and the Demands of Social Justice

Field observation tends to spark curiosity beyond the original questions brought by the researcher, and this is one strength of fieldwork. Because of the predicaments that researchers encounter in the field, humanists find themselves learning about nonhumans, even as natural scientists learn about humans. It is this aspect of fieldwork that might best address another concern of critical humanists: that scholarly attention to nonhumans will dilute the social justice focus of humanism, perhaps even shifting responsibility for great crimes—genocide, enslavement, exploitation—away from human perpetrators to "natural forces." In fact, much Anthropocene description shows how human suffering is wrapped up with nonhumans. Illness, food security, land rights—these are just a few examples of human concerns that *require* attention to nonhumans together with humans. The idea that studying nonhumans necessarily neglects injustice is just not true.

The *Feral Atlas* project with which this book is in dialogue offers a variety of field reports that tackle questions of human suffering through better attention to nonhumans. Consider the work of African American anthropologist Paulla Ebron, who came to the study of nonhumans while studying regional identity formation in the Sea Islands and coastal Low Country of the southeastern United States.[8] She found she needed to understand how the making of disease environments shaped the experiences of enslaved people.

Taking another look at the famous drawing of the slave ship *Brookes* circulated by nineteenth-century abolitionists (Figure 10), Ebron focused her attention not just on the bodies of captives but also on the cramped space between them. Wondering what else lived in those slivers of "empty" space, she turned her attention to the rats, lice, mosquitoes, and disease organisms that made survival even more precarious. This, in turn, prompted her to learn more about the yellow fever–carrying *Aedes aegypti* mosquitoes that came across the Atlantic with European ships. The mosquitoes may have come from the Mediterranean or Africa—but the inhuman conditions of slave ships, in which bodies were packed with so little regard for their welfare, allowed a consequential development: A new strain of mosquitoes evolved on the ships. African *Aedes aegypti* mosquitoes live in holes in trees, and they rarely encounter humans. But this new strain lives exclusively on human water sources, such as the fresh water carried in the slave ships. All the *Aedes aegypti* mosquitoes in the New World are descendants of the strain that evolved on slave ships. These human-loving mosquitoes infected enslaved people as well as their captors, shaping generations of precarious life and untimely death in the New World. In time, this slave ship lineage was brought back to Africa through ships, and also across the Pacific to Asia. This strain of mosquitoes, born with the suffering of enslaved people, continues to carry some of the worst viruses of our times, from dengue fever to zika.

Ebron's field report illustrates how humanist research can lead to attention to nonhumans without losing track of social justice concerns. In this work and that of many others, our ability to trace histories of human oppression and suffering—or really any human experience—depends on research focused on the relations between humans and nonhumans. Ebron's case also shows the importance of working at multiple scales—from the humid spaces between captured *bodies* as blood-meal sites for breeding mosquitoes to the shipping connections

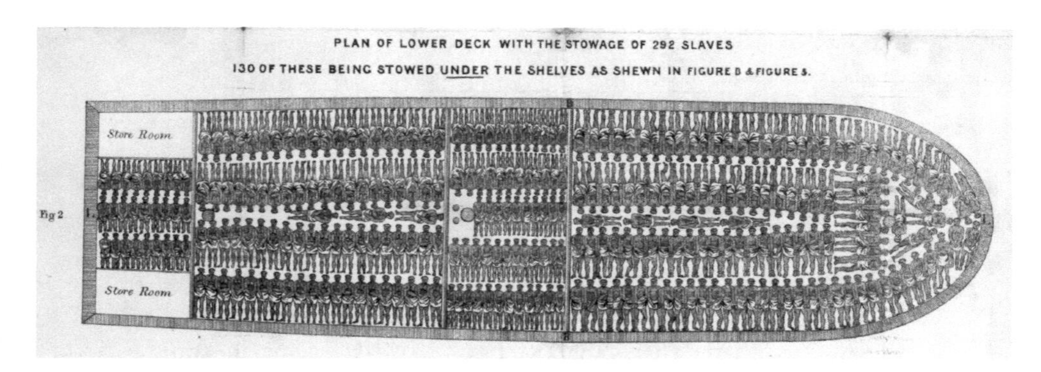

FIGURE 10. **The plan for the slave ship *Brookes*.**
UNITED STATES LIBRARY OF CONGRESS

between *continents* as channels for traveling viruses: It is only by working across these scales that it becomes possible to understand how disease environments are created. Indeed, this brings us to the globe, and even the planet. A number of scholars have argued that Atlantic slavery was central to the emergence of a global, world-making "modernity."[9] The making of anthropogenic disease environments is a crucial feature of this planet-changing modernity, and Ebron's research makes this explicit.

Critics worry that attention to nonhumans might take away from social justice activism, yet research on the more-than-human Anthropocene often emerges from collaborations between activists and scholars. Filipina American anthropologist Alyssa Paredes offers such research.[10] Paredes investigates conditions in a banana plantation in Mindanao, where those living nearby complain about the effects of fungicide spray, which drifts into their homes and gardens, wilting plants and sickening children as well as animals. Paredes worked with a social movement called Mamamayan Ayaw sa Aerial Spray ("Citizens Against Aerial Spray") to learn about the situation. Without the opportunity to work with the social movement, Paredes would not have been able to get to the thorniest part of the problem: the specific cocktail of poisons sprayed from planes and drifting across villages and gardens. Because the combination was considered proprietary, no scientific research about just that combination could be used in investigations and trials. Demanding reform is made much more difficult because critics are unable to gather the appropriate evidence.

As in Ebron's research, human beings have become collateral damage to elite attempts to remake the world for their own advantage. In Philippine banana plantations, toxic chemicals are feral; they are Anthropocene co-creators, along with project directors who may not care about local residents, whether human or nonhuman. One of the most dangerously misleading aspects of the term "Anthropocene" is that it urges us to imagine humans in charge, not just of other people but of the whole planet. But even the most powerful humans are only able to make things happen in concert with nonhumans. Furthermore, such entanglements so often exceed designers' plans—especially when designers don't care particularly about feral effects. Paredes's research illuminates one of the insidious features of the more-than-human Anthropocene: Many harmful agents, human and nonhuman, transformatively mix in creating environmental dangers, while the protections of property law and policy work to ensure our ignorance about just what is going on.

Paredes's research, like Ebron's, also compels attention to scale. Might it be useful to see the modern world emerging from a plantation condition?[11] To ask this question involves imagining ways that racialized terraforming projects create systems effects that ricochet across the globe, altering planetary life. Such questions move out from field observations into thought experiments about working across scales. Our team argues that scale-shifting experiments need not erase field observation; at their best, they make field-based knowledge more important than ever.

Shifting Scales in the Environmental Humanities

Consider how some other environmental humanists work across scales. A particularly clear example of working across scales can be found in Amitav Ghosh's *The Nutmeg's Curse*.[12] Ghosh aspires to sketch out a global regime—an "Imperiocene" perhaps—in which Europe's colonial expansion structures the conjoined social and environmental catastrophes of our times. But he begins in a particular time and place: 1621, in tiny Banda Island in today's eastern Indonesia, where the Dutch massacred the population to lay claim to the island's nutmeg trees. Having done so in the most brutal possible manner, they brought in enslaved labor for the production of nutmeg. Ghosh connects this history to the massacre of the North American Pequot at Mystic, Connecticut, in 1637, only a few years later. These two histories are his way to link European conquest of the Indian Ocean with the New World, creating a global story, with planetary implications. The two massacres, as particular as they are, set a pattern for what follows, Ghosh argues. They invite a recognition of the ways that distinct histories fit together across regions to create systems effects that encompass the world.

In Ghosh's thought experiment, the common thread tying the Indian Ocean and the New World is the logic of European governance projects, which strip the land of people, names, and more-than-human livelihoods, replacing them with imported European versions. These projects can be compared to the terraforming of other planets discussed in science fiction—and later in this book. Empire building has changed the world by wiping out the old and inserting the new—without regard to the welfare of either humans or nonhumans. Dredging harbors for imperial shipping is as much a part of this governance project as introducing enslaved labor.[13]

The stories that emerge from enslavement inspire the thought experiment of philosopher Sylvia Wynter, who characterizes the Caribbean world made by European conquest as the world of the plantation.[14] The plantation is native land grabbed and reshaped for European governance (as for Ghosh), but also (for Wynter) remade in its orientation to capitalist markets. Wynter adds to this plantation world the "plot"—that is, the garden given to enslaved workers to raise their own food for subsistence. Wynter argues that the plot is the place from which stories are generated: The plot of stories emerges from relations to the land in the garden plot. Caribbean storytelling comes to life at the intersection of plot and plantation, in both their contrasts and their interpolations. This too is a scale-shifting method for moving from particular stories to their systems effects; plots and plantations are refracted across multiple scales.

Wynter has also embraced the emancipatory potential of the planetary scale, as imagined for the Anthropocene.[15] She situates this in a world history in which Black people have rarely been considered properly human. Every "ecumenical" statement about humanity—that is, every statement that claims to create unity across differences—has been a lie, she argues: Only certain people are given the privilege of entering humanity. This makes claims about the human condition suspect. Yet, despite this, there is something particularly interesting for Wynter about the Anthropocene, with its planetary reach and its reference to a common humanity. The Anthropocene forces current philosophies and technologies to admit their failure. Appeals to the rationality of white bourgeois elites make no dent; instead, they show themselves as the problem. In this context, the planetary scale of the Anthropocene, according to Wynter, has a liberatory potential. Like Black critique, the Anthropocene, in its planetary danger, shows that everything that has preceded it is just wrong. Perhaps mobilization at this scale might allow, for the first time, a humanity that encompasses all. But note that, for Wynter, the planetary scale is just one way to mobilize; it moves into this role together with the plot and the plantation. In contrast to Chakrabarty's advocacy for the planetary scale, which argues for its exclusive nature, Wynter's planetary is an activist experiment for mobilization.

Without these scale-shifting stories, it is too easy to naturalize human action. The problem is particularly easy to see in discussion of the Great Acceleration, the period since 1945 that is currently the favored choice for geological certification as the Anthropocene epoch of planetary history.[16] Because the Great Acceleration is commonly imagined through a series of graphs that quantify human disturbance, the structural and systemic features of human–nonhuman relations are obscured. Without a structural account, programs of multispecies invasion, imperial governance, and capitalist expansion are naturalized as the only way humans can live on earth. Yet it is also perfectly possible to view the Great Acceleration with a different narrative, as a development in political economy and ecology (see Chapter 6).[17] One might argue that Rachel Carson's *Silent Spring*—with its attention to the normalization of toxic chemicals as part of the post-WWII "peacetime" economy—was an early form of just this kind of political ecology.[18] To explain structural violence against more-than-human livability, we need attention to agendas of landscape modification, which in turn rely on field observations. From field observations, it is possible to move to thought experiments that show us systems effects at multiple scales.

Global climate modeling is a scale-shifting thought experiment that deserves the same kind of critical attention as the experiments of the philosophers and historians discussed above. The key is not to let the model allow us to forget the field research and scale-shifting experiments on which it is based. We need the planetary scale—as one among many that ultimately depend on the work of field observation.

Multiple scales are necessary not just for our study but for our response to the Anthropocene. Global mobilization against climate change is a fine goal, but it can hardly stand alone. Many Anthropocene effects need action at other scales. Consider the spread of plant diseases through industrial commerce. Many planetary modelers tackle the impact of climate change on forests, but the emissions-reducing actions advocated to stall climate change rarely encompass the kinds of regulatory actions needed to halt the importation of plant diseases, which are likely to destroy forests long before global warming gets to them (see Chapter 9). Plant diseases are ignored by these modelers because their traffic needs to be studied—and responded to—at the scales of industrial supply chains and shipping policies and routes, rather than at a unitary planetary scale.

Bitty Roy and her colleagues take on this problem.[19] Their research describes how forests are dying from industrial shipping practices that allow disease organisms to be moved from continent to continent every single day. Forests can withstand a certain level of disease. But the speed and scale of the importation of plant diseases today is unprecedented, and forests, new to these imported threats, are unable to muster resources against them in a timely way. Even as Anthropocene policymakers imagine planting forests as a means to sequester carbon, they refuse to regulate industrial shipping as it transmits forest-destroying diseases.

Roy and her team are natural scientists and science policy experts; they have practical goals in mind. They suggest kinds of regulation that might address this problem. Their suggestions are international—but not planetary in the senses either Chakrabarty or climate modelers have in mind. They focus on what ought to happen at ports of entry involving the regulation of wood products and the shipment of live materials. This is just one example of an important Anthropocene phenomenon that needs to be addressed at an other-than-planetary scale. At the same time, there are clear planetary implications for the death of forests, including to atmospheric carbon dioxide levels. Anthropocene threats may operate, then, at multiple scales; and both research and response need to follow across those scales.

Making Field Observation Matter to Anthropocene Studies

If we take field observation seriously, a number of analytic tools suggest themselves for studying the Anthropocene. The first step is to take the patch as a research object. Patches cannot be determined from a top-down plan; they emerge *from* field observation—through the contours of the phenomena being studied. Anthropocene patches, in the landscape ecology sense, include parking lots and plastic gyres; unlike Wynter's "plots," these are not mainly sites of utopian possibility. Instead, they show us the Anthropocene, in all its horrors. Plantations are Anthropocene patches.

Scholars cannot afford to naturalize these patches; instead, field description is necessary to better understand what is happening in them and how they came to be. As an aspect of this description, we must consider the importance of imperial and industrial infrastructures, which modify landscapes, making Anthropocene patches. We must also consider what kinds of maps and mapping practices best capture and convey Anthropocene phenomena. We turn to these questions in the next chapters.

TWO

What Makes a Patch?

The concept of the patch has mainly been used to map landscape ecologies in which humans have had little influence, such as forests and coral reefs; however, some authors have attempted to map Anthropocene geographies in relation to the ways humans have engineered varied spaces, such as cities, villages, and croplands.[1] This book updates these projects through attention to feral geographies: stimulated by human infrastructures but outside human control.

Patches can be identified at many scales, from a spot on a leaf to an ocean. The scale for any research project depends on the focus of the research; in choosing the correct scale, the researcher follows the phenomenon. The examples below should make this clearer. But before moving to examples, it seems important to say that, as an analytic tool, identifying patches cannot do everything. Identifying a patch does not reveal its social and ecological dynamics; that is a separate step. Still, it's a useful beginning in moving away from the homogeneous planetary space that possesses the popular imagination *as* the Anthropocene—and inspires many scholarly reactions, from the embrace of too-big-to-imagine "hyperobjects" to the condemnation of world-systems theorists.[2] A patchy Anthropocene is spatially heterogeneous, and this heterogeneity should be the object of our attention. Patches are a place to begin. Furthermore, to establish spatial heterogeneity, we can include ecological corridors—that is, lines along which organisms or nonliving things move—as a particularly linear form of the patch.[3]

Including corridors should not push us to de-emphasize places in favor of nonmaterial networks, such as those that connect experts and the objects of their expertise. Nor should we abandon places for discursive formations, that is, the clusters of participants, institutions, and ideologies that, as many social scientists have usefully shown, give power to places and the beings that happen to be there. Instead, we begin with the landscape ecology definition of the assemblage—as that collection of beings that happen to be in the patch—whatever their relations with each other and with humans. This kind of patch is usefully open-ended. It spurs our curiosity to ask, What's going on there?[4]

The advantages of designing research through the patch are most obvious when patch boundaries are rather definitive, as in a mainly contained body of water, such as the Black Sea. The Black Sea has outlets only at the Kerch and Bosporus Straits, on its north and southwest sides, respectively, as shown in Figure 11. The map shows currents within the Black Sea, which circulate water within the patch, with little interference from other bodies of water. These are the conditions under which the Black Sea became host to a great bloom of comb jellies in the late 1980s, as studied by the lab of Turkish marine biologist Temel Oğuz and reported by lab members Bettina Fach and Baris Salihoglu.[5] By 1989, they found, there was a "regime shift" in the organisms in the Black Sea. Instead of fish, there were only jellies. The jellies ate the fish when they were still tiny; the fish could not replace themselves.

What had happened? The containment of the sea is the key to their story. Once the Black Sea was a great fishery, but by the 1980s, the big fish had been removed through overfishing, and the fisheries concentrated only on anchovies. Meanwhile, runoff from commercial agriculture swept too many nutrients into the sea, causing eutrophication, which killed fish. With the amount of fishing as well as the deterioration of water quality, the fish were already in trouble. When comb jellies were accidentally introduced in the ballast water of incoming ships, and in the context of particular weather conditions, the jellies were able to take over—completely changing the system from dominance by fish life to dominance by jelly life. Fortuitously, a different jelly organism was accidentally introduced into the sea in the 1990s, and the new jelly began

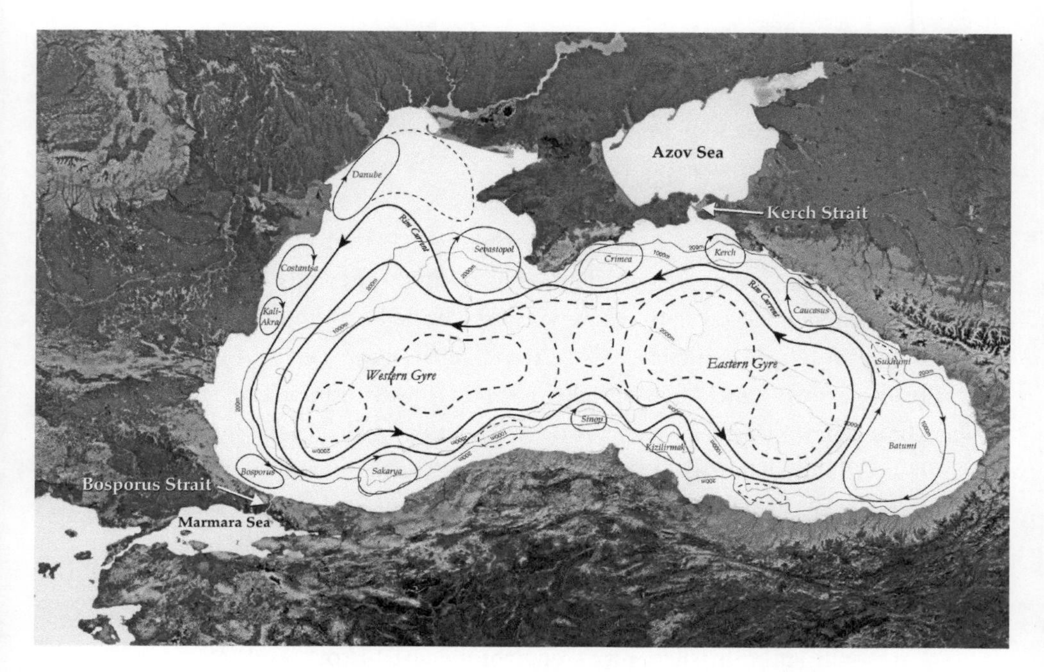

FIGURE 11. **The Black Sea is a deep, landlocked basin with exchange only to the Marmara Sea through the Bosporus Strait and to the Azov Sea via the Kerch Strait.**

LILI CARR

eating the original one. This allowed for at least some amount of biodiversity to return to the Black Sea.

The Black Sea has been, as the researchers put it, a "natural laboratory" for studying jelly invasion. The containment of the water meant that fish were not being replenished from the ocean, that excess nutrients were not being washed out to sea, and that when an exotic organism was introduced, the entire sea was its field of operation—allowing a complete takeover of the ecology. Obviously, there are connections that need to be considered, including the lack of regulation of interocean shipping, resulting in ballast waters full of potentially dangerous organisms, as well as the lack of regulation of agricultural runoff from surrounding lands. But in this contained space, the patch shows itself as a powerful tool for analysis.

Identifying the patch has been a key starting point in environmental justice research. Attention to the placement of industrial toxins, whether dumped or remaining from abandoned industrial sites, has allowed scholars and advocates to establish and confront unjust racial ecologies in which people of color are asked to bear the burdens of industrial poisoning. Identifying these toxic sites—as patches—is the first step.

Scott Frickel offers a study of brownfields—that is, abandoned industrial sites—many of which release toxins, especially in storms. In one instance, Frickel maps these sites in Rhode Island, and especially in its capital city Providence, where toxic substances seep and ooze from dozens of brownfields. As Frickel explains, "A fundamental feature of urban economies everywhere, industrial churning drives the incremental yet continuous spread of chemical hazards throughout urban areas . . . altering the chemical composition of city soils, water, and air."[6] Identifying brownfields is a descriptive task, necessary to understand cities. But there also is an ethical dimension. Frickel continues, "Recovering the lost history of these sites . . . gives city residents opportunities to reclaim a personal stake in their city's environmental history and its future." This is an issue for advocates as well as scholars. Social justice demands that we investigate unequal exposure to industrial toxins.

Environmental justice must be inside, not an add-on to, Anthropocene studies—a note that has recently been underscored by the call to address the uneven distribution of the impacts of climate change in terms of "critical climate justice."[7] Working with patches requires that we differentiate across ecologies—including those implicated in injustice. It is an important first step to understanding how landscapes are shaped into racial ecologies.

Human Infrastructures Stimulate Patches of Feral Activity

To recap, in this book, infrastructures are projects that modify the land, the water, and the air. Infrastructures have feral effects, and these compose the Anthropocene.

In anthropology and science studies, infrastructures have most commonly been defined as networks that allow people, goods, and ideas to travel.[8] This allows an analysis of all the kinds of people that use a network, including those who don't like how the network functions.[9] Through network-based theoretical work, the literature on infrastructures has usefully illuminated emergent political aspirations, dreams, and communities.

This framework is great for asking some questions, but not so great for others. We will return to political visions in Chapter 4, in asking why imperial and industrial infrastructures are built in repeating forms in so many places. Here, however, we need to see feral effects, which are obscured by this analytic framework. We need a more material definition.

Network analysis of infrastructure can attend to materiality—but only as it relates to the working of the network. This is because networks, which define infrastructure in such approaches, are not in themselves material; materiality must re-enter the analysis as it contributes to the network. Leakiness in the water system is inside the water-use network, but mosses living in the same water system (if not annoying to human water users) are not. The mosses may be part of a different network, but not the one defined by human water users. To take material arrangements outside the network seriously, it becomes necessary to return to the definition of infrastructure, to start with its material arrangements and forms, rather than its uses.

Feral effects are visible only when infrastructures are studied in their materiality—and as they make social and ecological patches and assemblages. As in network-oriented analyses, the material effects we chart are relational, but these relations are not confined by the designed function of the infrastructure and those who participate in it. For this work, then, infrastructures are building projects, not networks of users. In general, this is the definition used throughout this book.

The form and material features of infrastructure matter. Marine infrastructure, for example, provides underwater flat surfaces that are rare in nature. Jellyfish polyps, as studied by Martin Vodopivec and his lab, love these flat surfaces, forming great colonies through cloning, which mature into medusas that float across marine space. Marine sprawl increases the jellyfish population to boom—just because of its form.[10]

The feral effects of human infrastructure can be benign and ordinary. Zoologist Peter Funch writes about the meiofauna—tiny creatures—that live in urban drinking water, despite chlorination.[11] Bacteria are killed by the application of chlorine, but not these animals. Luckily, most people are not hurt by drinking and digesting them. But their presence is an example of nonhuman uses of infrastructure. Meiofauna would be at best peripheral participants in a network analysis of the water infrastructure (perhaps through the complaints of water users). In a material analysis beyond the users' network, they can take center stage. Their plentiful presence in human drinking water systems is an example of the infrastructure's feral effects—and a reason to focus on the material dynamics of infrastructure, as it makes patches for the livability of some beings but not others.[12]

The patch formation of meiofauna in urban water systems is a feral effect; the same array of species outside of the city might also form patches, but not patches of ferality. Thus, too, for invasive ants in California's cities: Ant biologist Deborah Gordon explains that invasive Argentine ants love people's houses, where there is just the kind of food and water they like.[13] But in wilder settings, native California ants fight back; the dominion of Argentine ants is far from guaranteed. Argentine ants hitchhiked on ships; they are encouraged by human housing. Their engagement with ships and houses creates their feral behavior; they are nonhuman users of human infrastructures. Thus, too, these ants turn out to be less of an ecological problem than imagined; they do not spread everywhere. Only in their relations to human infrastructures do these ants form Anthropocene patches.

Some feral effects are more dramatic, even in their patchiness. Staying with urban water, consider the arrival of rats in sewage systems, as described by historian Michael Vann.[14] Rats are not part of the intended purpose of sewage systems, but it turns out that rats flourish in the humid and nutrient-rich spaces of urban sewers.

Vann's research concerns the colonial construction of Hanoi in French Indochina. As in their other colonies, the French cleared space to build a beautiful modern city for white residents, while leaving the crowded remains to colonized Vietnamese residents. One of the hallmarks of the modern city was its European-style bathrooms, connected to an extensive sewage system. It was to be the pride of the colonial occupation, showcasing the wonders of civilization. Then the rats came. The modern sewers turned out to be an excellent habitat for rats, which thrived and multiplied. Crawling through the pipes, they emerged into the modern bathrooms of the French colonizers (Figure 12). The French were horrified, and they mounted campaigns in which

FIGURE 12.
LIZ CLARKE

Hanoi residents were paid to bring rat tails to colonial authorities. Imagine their surprise when tailless rats began to crawl out of the sewers! The news that rats were even more prevalent in the white city than they were in the native city, because of the former's sewage system, was a blow to the pretensions of French colonialism. Like so many feral effects, rats in the sewers belie the utopian claims of modernization and development.[15] Vann's history opens the question of how European colonial expansion created a suite of feral effects, a question considered throughout the rest of this book. Here it seems useful to note that utopianism itself encourages feral effects by ignoring and overwriting the presence of local communities, human and nonhuman.[16]

Feral Effects of the Plantation Form

One infrastructural project that tends to have dramatic feral effects is the monocrop commercial farm, the plantation. While the term "plantation" is the ordinary term for commercial tree-crop cultivation in the contemporary world, it also harkens back to the legacy of the European cultivation of exotic commercial crops around the world using coerced labor—including the New World's plantations using enslaved African people.[17]

This legacy of violence is useful to understand the tree-crop plantation today. The earlier plantation conditions helped create taken-for-granted features of the modern world, such as the discipline of labor to produce profits for distant investors.[18] The use of monocrops emerged in part from planters' counterfactual idea that the work could thus be done without much attention to the farming process. In fact, the enslaved people working the farms had considerable knowledge about plants, and this helped them survive under difficult circumstances.[19] Still, under plantation conditions, workers have often developed a hostile relation to plantation crops. Sidney Mintz, for example, explains that sugarcane workers in Puerto Rico in the 1950s described themselves as "going to battle" with the cane.[20] Because of the speed and discipline plantation owners demanded for cutting, the cane did, indeed, cause numerous worker injuries. Similar problems of alienation and injury are reported in contemporary tree plantations. Tania Li and Pujo Semedi found that oil palm plantations in West Kalimantan, Indonesia, were established with a kind of racialized discipline in which workers were sacrificed to efficient production.[21] Furthermore, contemporary commercial tree plantations continue to be associated with the violence of dispossession, as Indigenous residents are removed from their traditional territories.[22] The term "plantation" has returned, indeed, to describe contemporary land grabs, involving all kinds of crops.[23]

It is useful to reproduce Li and Semedi's respective definitions of the plantation—as exemplified in Indonesia—to get a sense of the term's contemporary traction.[24]

Pujo: A plantation is a giant, an inefficient and lazy giant, but still a giant. It takes up a huge amount of space. It is greedy and careless, destroying everything around. It is alien, strange, and unpredictable. It is human, but you cannot form a normal human relationship with it. It can trample you, eat you, or drain your strength then spit you out. It guards its treasure. You cannot tame it or make it go away. You have to live with it. But it is a bit stupid, so if you are clever you can steal from it.

Tania: A plantation is a machine that assembles land, labor, and capital in huge quantities to produce monocrops for a world market. It is intrinsically colonial, based on the assumption that the people on the spot are incapable of efficient production. It takes life under control: space, time, flora, fauna, water, chemicals, people. It is owned by a corporation and run by managers along bureaucratic lines.

Through its numerous legacies of violence, the plantation thus straddles and connects imperial and industrial legacies. In the process it creates dangerous Anthropocene patches. Chapter 1 discussed toxic chemicals on Philippine banana plantations, which threaten local people, their animals, and their crops. Monocrops also attract and nurture diseases, which can spread far beyond the plantation to destroy what otherwise might be much more sustainable peasant farms.

Agroecologist Ivette Perfecto has studied this problem in the coffee plantations of Central America.[25] It's possible to grow coffee under varied conditions, as it does quite well in the shade of other trees. Peasants in Central America have a history of growing coffee together with other tree crops in polyculture food orchards. However, in the last few decades, monoculture, full-sun coffee plantations have also made an appearance. Figure 13 shows the varied ways coffee can be grown in this area, which range from allowing coffee to be an element in an anthropogenic forest (as shown at the top of the figure) to full-sun monoculture (at the bottom).

It turns out that these varied methods of growing coffee have very different effects in relation to coffee diseases. Coffee rust fungus is a devastating coffee disease that is spread by wind-borne spores. Planting coffee in the shade of other trees offers considerable protection from coffee rust fungus. In contrast, full-sun monoculture allows spores to gather, launching an attack of the disease. Furthermore, once the disease is established in a particular place, it becomes a source for infection across the whole region, creating an ever-growing patch of infection. At some point, the disease becomes so saturated in the region that it can no longer be controlled; it infects peasant polyculture farms as much as monocrop plantations. At this point, it becomes an epidemic. No trees are safe. Furthermore, just planting more shade trees no longer solves the problem. This is a "state change": a transformation of the ecological dynamics of the region. Perfecto explains the problem through the concept of hysteresis. Once a state change has occurred, it becomes very difficult to go back to the original state. Coffee may no longer be a viable crop.[26]

As mentioned in the Introduction, this book singles out imperial and industrial infrastructures, rather than all human infrastructures, to explain the Anthropocene. We argue that the scale of imperial and industrial infrastructure building in the last five hundred years has caused more systems ruptures—that is, state changes—than ever before in human history. The concept of state change suggests how deeply these infrastructures can affect landscapes. The whole matrix of peasant food crops and agroforestry as well as their cash crops are affected; a new patch structure is coming into being. This poses problems for understanding time as well as space. The discussion of state changes will be continued in Chapter 4.

What are the best ways to illustrate the feral dynamics of Anthro-

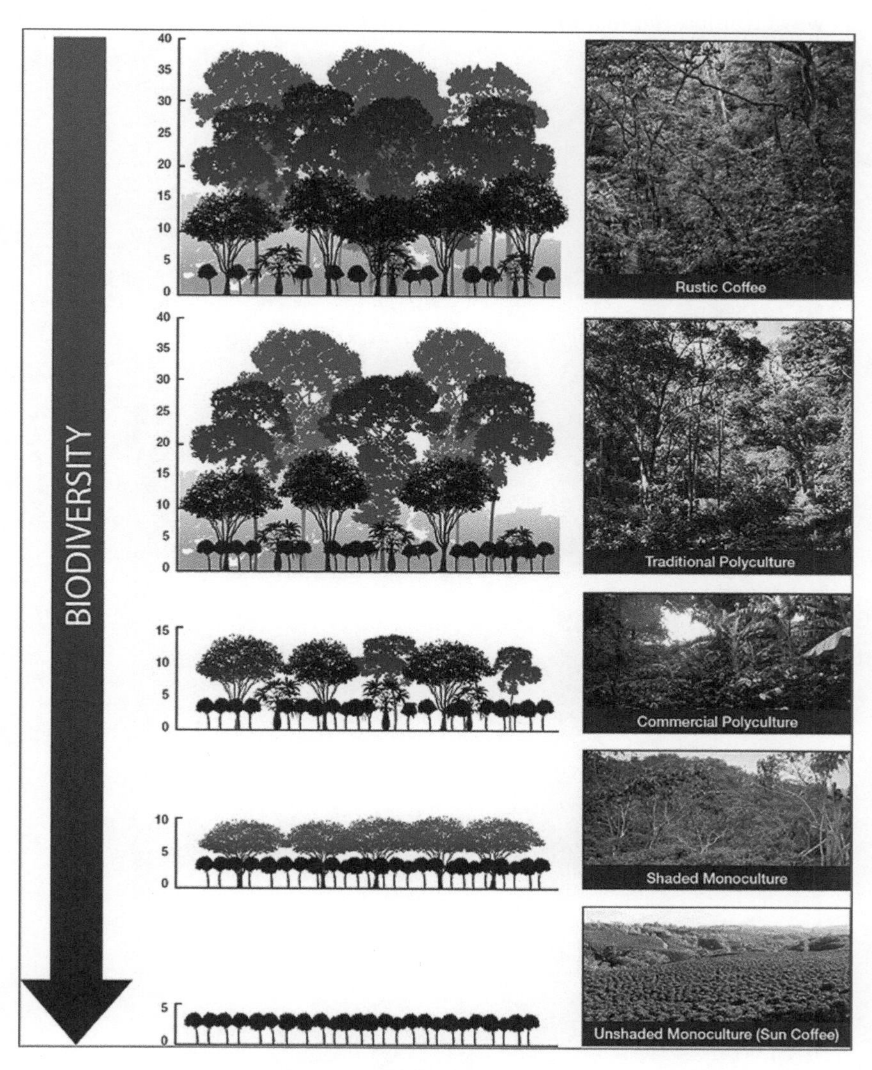

FIGURE 13. **Coffee intensification (technification) gradient.**
DAVE BRENNER

pocene patches? This book has already offered a variety of images and diagrams as aids to explaining patch-based feral effects. Only a few are easily identifiable as "maps." Yet each aids our ability to analyze the patch as a site of feral action. We now turn to further discussion of maps as ways to investigate and describe Anthropocene patches.

THREE

Mapping the Patch

In recent years critical cartographers and Indigenous counter-mappers have made an important point: Maps don't simply show the world as it is; maps create specific kinds of territories—and specific ways of moving through the world. For centuries the maps of empire and capital have worked to contain, domesticate, distort, and erase territories on a global scale.[1]

Developments in computational mapping have, in turn, enhanced and, we would argue, emboldened this presumptive force. With GIS enabling ever-more-sophisticated techniques of data capture and visualization, western mapping traditions have claimed an enhanced empirical, analytical, and predictive purchase on the world, at a grand scale (significantly influencing research funding and agendas). As maps make visible social and material arrangements of people, places, histories, and relations, they express particular orientations to the past, present, and potential futures.

Because maps teach us how to move through the world, they shape what we attend to, what we avoid, and what we overlook altogether. In making certain things visible and accessible, they reinforce particular realities and displace others. Maps, in other words, can have a profound impact on how worlds are arranged and sustained, domesticated, and contained. Or made to disappear.[2]

This chapter embraces mapping as a means of insisting on a multiplicity of perspectives, scales, and representational practices. Where modernist cartographic traditions employ a more or less universal vocabulary in map-making—a single projection (typically a plan view); a defined scale; a set of line weights; and the use of textures, colors, and shadings for information hierarchy—we push elsewhere. For some kinds of information cannot be translated through data visualization, and some spatial qualities cannot be translated merely by lines or color shadings.

If colonial and modernist mappers envisage space as empty—awaiting occupation, order, and the imposition of authority—the task for critical Anthropocene cartographers is to highlight the lethal consequences of such perspectives. In *Feral Atlas*—a collaborative project involving scientists, scholars, artists, poets, and vernacular and Indigenous observers of the Anthropocene—we curated and created a series of multimedia maps to spatially depict the patches described within each field report. Nothing in these maps is fixed or manageable. Our maps expose not only the grandiose illusions of mastery that have gotten us into this mess, but the cascades of consequence that follow.

Unlike speculative mapping projects that critique colonialism and capitalism through imaginary landscapes or conceptual diagrams, our maps remain analytically committed to depicting grounded material relations. They do so by identifying specifically located trajectories of breach, rupture, and spill: depicting ecological patches disrupted and refigured by the world-ripping force of feral entities on the move.[3] This provides the grounds for a critique of imperial and industrial infrastructure projects arising directly out of the material relations under examination.

Assembling Patchy Perspectives

The five-hundred-year time span of our project corresponds, unsurprisingly, with the history of the modern Western tradition of cartographic mapping, though a more diverse set of mapping traditions, some of which we draw from, have existed for millennia.[4] Pākehā and Māori researchers Alison Jones and Kuni Jenkins forcefully identify mapping's ongoing role in colonial world-making:

> The modernist project of mapping the world . . . is an expression of a Western Enlightenment desire for coherence, authorization, and control. . . . The assumption held by dominant cultures . . . that everything

is in principle knowable . . . [fuels] fantasies and desires for a better, less fragmented world.[5]

These days ubiquitous map apps geolocate and direct anyone with a screen device and Wi-Fi connection, cultivating new forms of insular (in)attention to the material, ecological worlds being made and unmade around us. Google Earth and Google Maps have become taken for granted—as enormously useful, if not deeply pleasurable, for billions of users across the planet. Google Earth renders the "God View" interactive: Anyone with a phone and Wi-Fi connection is free to roam, zoom, and survey once distant and inaccessible places from a lofty, and comfortably distant, perspective within the practical limits of accessible data sets and download speeds.

Even as these layered, zoomable digital maps expand the ways we access spatial information, they direct our attention and so cement a certain version of what is sensed, what is known, and what remains overlooked. Every time we use a Google Map, we participate in the ongoing making of the world as a project of modernity and by extension become complicit in the erasure of other worlds and other ways of knowing, seeing, and caring about the world. Such maps introduce forms of hierarchy, bias, and distortion, which get translated into our own understanding of space as we use them. For instance, the Mercator projection, widely displayed in twentieth-century classrooms and textbooks, and now the template for most digital mapping, embeds a deeply biased, and as many have argued, colonial perspective into everyday ways of seeing. It does so by distorting the relative size of landmasses, shrinking Africa and South America, and enlarging northern regions including Europe and North America, for the sake of a flat and rectangular depiction. Meanwhile, our smartphones turn us into active mappers, consciously or otherwise. When we share a dropped pin, enable others to track our movements, or upload geolocated images to social media, we are being recruited into the reproduction and re-instantiation of the highly specific, selective, and technologically enabled worldview embedded in technologies developed and distributed by giant, data hungry, profit-seeking corporations.[6]

This is the perspective that our maps in *Feral Atlas* set out to destabilize and expand.[7] While remote sensing and associated technologies remain key to the ongoing development of climate science and many other important forms of environmental studies, our maps provide a critical counterpoint to big data, planetary perspectives. With a focus on situated material relations, and the art of attending to temporal as well as spatial relationships, the diverse mapping practices within *Feral Atlas* argue for a critical appreciation of maps that don't easily scale up or down, much less fit together to form a coherent, homogeneous, and fixed-in-place whole.

At the same time, these maps do more than represent and describe. As they work to destabilize and overthrow modernist certainties, they guide us ever deeper into the patchy Anthropocene and towards the kinds of knowledge work that we argue is essential if we are to better research and report on shared worlds under threat. Throughout this book when we talk about the work of description and representation, we do so with a sense of the world-making force of such knowledge practices.

Mapping Anthropocene Patches

Place is central. Because we are studying the Anthropocene through its patches, our analyses are place-based, which is one reason we call our digital project an atlas.[8] However, pinning a feral effect to a spot on a world map is only occasionally the best way to understand a patch. To better convey the dynamics of Anthropocene patch-making, our maps are also composed of diagrams, images, collages, animation, and even sound recordings that allow us to address questions such as these: How did the feral effect develop and spread? How was infrastructure implicated? What gaps and absences has this feral process left in its wake?

Consider the spread of the American bullfrog, as discussed by natural scientists Nathan Snow and Gary Witmer.[9] American bullfrogs have been introduced in many places around the world, particularly in Asian and Latin American development projects that aimed to provide cheap, if not particularly desirable, meat for the poor. Yet American bullfrogs are resourceful; it is rarely possible to keep them contained in the enclosures built for them. They jump out, and spread across the countryside, eating native frogs and spreading diseases that kill off whole populations, while their own group remains relatively unscathed.

What might be the best way to represent the spread of introduced American bullfrogs? Our team scoured the scientific literature, hoping that some researchers might have mapped the trajectories of bullfrog escapes. While we found discussion of the problem of frogs escaping, none offered a representation that might help readers understand the relationship between building enclosures for introduced frogs and the spread of frogs and diseases across the countryside.

In the absence of such maps, *Feral Atlas* visual director Victoria Baskin Coffey created an image (Figure 14) that shows frogs in the process of escaping. Her watercolor painting shows the ineffectiveness of enclosure for this species. Note the enclosure facilities that are sketched in the background of the image; they show two levels of containing walls that, even combined, do little to stop the frogs' escape. To further en-

FIGURE 14. **American bullfrogs escape from pens.**
VICTORIA BASKIN COFFEY

hance our understanding of what happens once the frogs escape, we paired her image with a photograph taken by a researcher in the lab of Brazilian frog researcher Fabricio Oda (Figure 15). The photograph shows a free-living American bullfrog caught in the act of swallowing a native Brazilian frog. Its voraciousness adds to the story of its escape act by illuminating feral processes and their effects—including the disappearance of native frogs, with consequences for stream ecologies across the region. Through such mappings, the spread of American bullfrogs and their impact on the environment make sense not as geographic points but as processes.

Sometimes feral spread takes place across scenes and scales that cannot be usefully captured by still images. We pondered how to best show the process of the development of antibiotic-resistant bacteria, as described by medical anthropologist Jens Seeberg.[10] His report in *Feral Atlas* describes two distinct geographic locations in which environmental saturation with antibiotics has resulted in the development of resistant strains: around pharmaceutical plants in India, where antibiotics

FIGURE 15. **Adult American bullfrog (*Lithobates catesbeianus*) swallowing an adult Chaco tree frog (*Boana raniceps*).**
EDUARDO GROU

seep out with wastewater, and in the commercial pig farms of Denmark, where antibiotics are used when taking piglets from their mothers before they have developed their own immunity. These geographic locations are important to Seeberg's research, but they are not the only places in the world where bacteria are transforming in response to human attacks. Seeberg discusses this widespread process as it is likely to make antibiotics useless in the next fifty years—not just in India and Denmark but around the world.

Rather than focus on these two places, then, the map for Seeberg's report gets closer to the process itself. We chose an experimental setup developed by the Kishony Lab to show how bacteria develop ever-more-resistant strains when confronted with ever-higher doses of antibiotics. In the experiment we see a giant rectangular plate of growth medium in which the levels of antibiotics increase—marked as bands—as one moves toward the center (Figure 16). Bacteria placed at both ends advance toward the middle, but they are blocked by antibiotics as they reach each new vertical band of the plate. At each challenge, new resistant strains develop and break through. The Kishony Lab video (available on *Feral Atlas* with Seeberg's field report) shows the process whereby bacteria come to occupy almost the entire plate, despite increasingly higher concentrations of antibiotics.

This does the work of mapping by locating and illustrating the feral process in relation to the landscape challenges it confronts (in this case, the medium for growing bacteria, complete with antibiotics). In landscape architect Jill Desimini's terms, this map becomes a means to understand "the past spatial conditions of a given territory as well as its future potentials"; a way of articulating "the space between the past—what is no longer—the present—what is given—and the future—what is yet to come."[11] The landscape is reshaped every time a new mutation allows a strain of bacteria to grow on a higher level of antibiotic-infused medium. It is not hard to imagine the "natural" forms of this experiment in streams outside pharmaceutical factories or in the feces of industrially grown pigs. The map shows the development of feral bacteria, attuned to human antibiotic applications, but outside human control.

The point of going through these experiments in mapping is to consider what it means to describe an Anthropocene patch. In each of the reports we've been discussing, we consider attunements through which

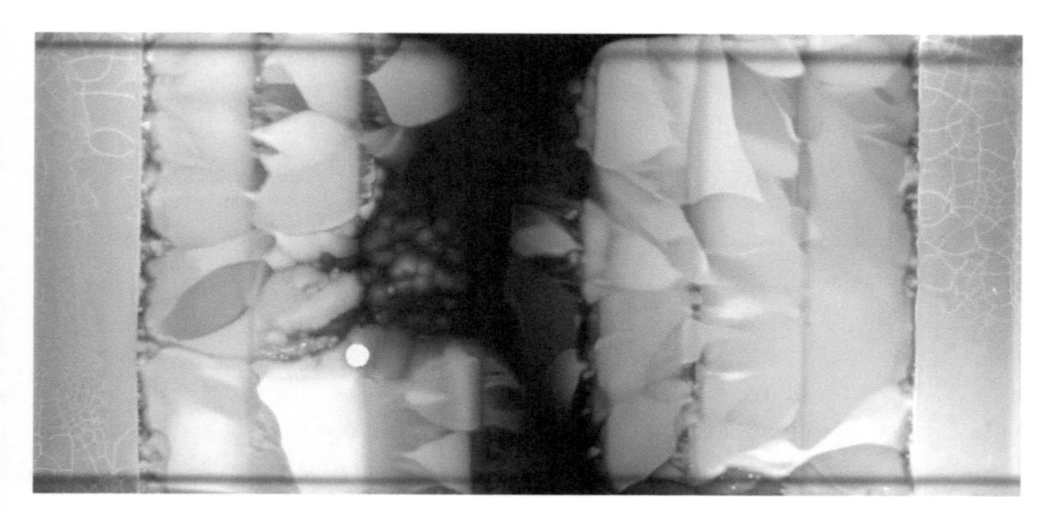

FIGURE 16. The video still shows bacterial strains that have adapted to move past the increasingly high-level dose of antibiotics of every band. White indicates bacterial growth; black is the basic medium without bacteria. Funnel-like shapes show a particular mutation that has developed the necessary resistance to break through to a higher antibiotic band and expand its population there; the small end is its starting point, and the large end its growth in the newly occupied zone.
KISHONY LAB

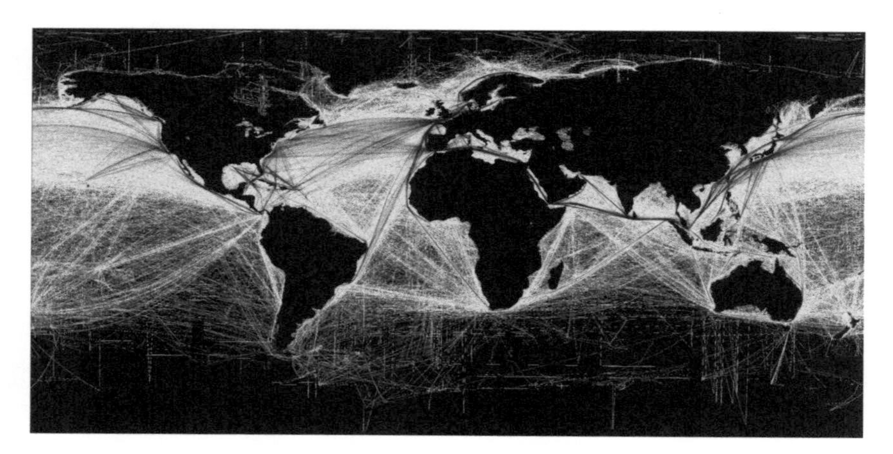

FIGURE 17. Cumulative commercial shipping movements in 2012.
ADAPTED FROM SHIPMAP.ORG

nonhumans adapt to living with imperial and industrial disturbance. American bullfrogs and antibiotic-resistant bacteria each become collaborators in building a dangerous Anthropocene through their feral adaptations.

Instead of focusing on the question of just how much territory these collaborations have conquered, the mapping practices we push show readers how to identify feral processes. These diagrams are intended as potential tools, including for the analysis of feral processes not represented in the atlas. When we stretch the concept of the map in these ways, we have a chance to move beyond geographic points to see environmental relations in motion—the key to a patch-based analysis.

Tracing Patch-Making Processes

Time plays a crucial role in our maps, and not simply in terms of identifying a specific origin event, or by allowing for the flow of time as it might be accounted for within a single map. These maps also gesture to aftermaths—indeed, a multiplicity of them.

Even as they depict a particular event, or set of circumstances, each map within the atlas can also be seen to be standing in for a multitude of others, unfolding in other spatiotemporal configurations: whether they are happening concurrently, or have already happened, or signal the threat of feral flows yet to arrive. Readers might consult, for instance, the video map of cruise ships entering and leaving St. Thomas harbor in Nathalia Brichet's field report on antifouling paint.[12] Or the map of international shipping routes (Figure 17) that introduces Marisa Weiss's report on the spread of emerald ash borers via shipping pallets.[13] In charting these various, and poorly recognized (beyond specialist fields) vectors of ferality, these maps disrupt not only cartographic conventions and the triumphant claims of progress and development that they embody. Grounded in the empirical specificity of past or present events, they depict a future that has become unfixed and unmanageable; a future currently being composed through infrastructure-mediated trajectories reshaping ecological relations across the planet: a threatened and threatening future.

Indeed, these maps undo a sense of the future occupying a singular timeline. They depict instead a multiplicity of Anthropocene temporal trajectories: a terrifying cascade of consequence set to arrive in different, overlapping, and multiple timelines, impacting some humans (to keep the frame here for a moment) more profoundly than others.

Our maps therefore depict time not as a dot on a timeline, but as marks and movements within landscapes that are in the process of being shaped by a catastrophic "meeting-up of histories" in Doreen Massey's terms,[14] or what we describe as the *piling up of histories until everything collapses*. Time is always in flow in our maps, at speeds and scales determined by the subject of the field report. Many of these maps gesture not just to spatiotemporal spread, but to a cumulative force of effects, to relational encounters and trans-patch temporalities beyond the borders of a single depiction.

One mapping experiment clarifies how feral patches empty out places as well as filling them up with Anthropocene collaborators across time. Agro-ecologist Marcela Cely-Santos studies pollinators in the Anolaima Mountains of Colombia.[15] Commercial agriculture, with its monocrops and toxic chemicals, has been harmful for insect pollinators, and Cely-Santos's research shows the disappearance of one insect after another from the valley. In response, *Feral Atlas* map director Lili Carr created a mode of mapping this disappearance by beginning with a screen full of the many kinds of bees that were once present in the valley, then showing them disappear, one after the other. (The order of disappearance is that observed by Cely-Santos in her research.) As each species disappears, it briefly leaves its name; as more disappearances occur, even the names are gone (Figure 18).

At the end of the short video that forms this map, only two species of pollinators are left on an otherwise blank screen. One is the Africanized honeybee, an imported species from the Old World. The other is a *Trigona* known locally as the dog dung bee because it builds its nests from feces. Both species do well in the simplified ecologies of commercial agriculture. Yet both are locally despised: The Africanized honeybee is very aggressive, and the dog dung bee destroys crops in its search for resins. Furthermore, as Cely-Santos points out, "these bees are a constant reminder of those pollinators that no longer fly through fields and forests."[16] People and crops are dependent on the pollination work of

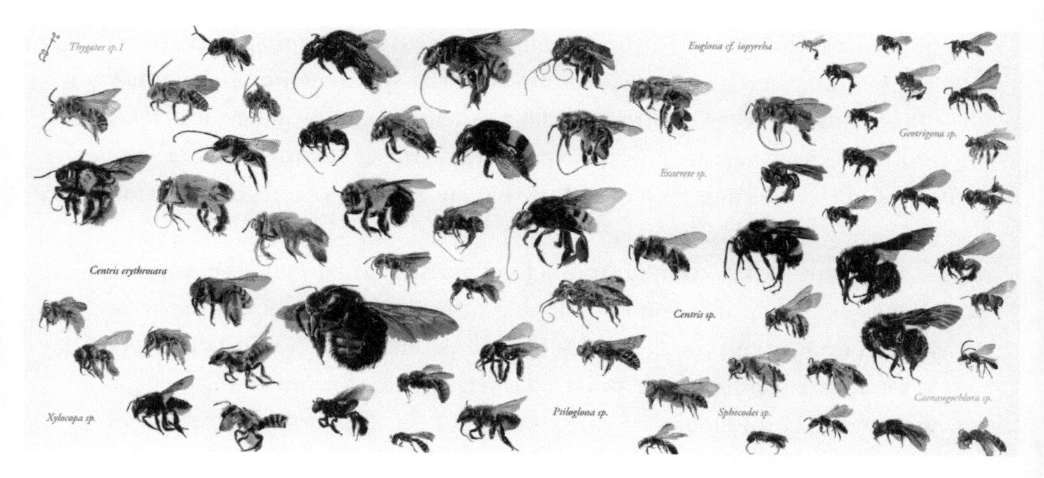

FIGURE 18. **Some of the more than seventy bee species found across mid-elevations in Anolaima, Colombia. Bees are depicted to scale relative to one another.**

MARCELA CELY-SANTOS, NICOLE DEGER-BEAUMAN, AND LILI CARR

these two bees. Yet, as Anthropocene collaborators, they carry the scent of the absence of the many pollinators that have disappeared.

A specifically located and empirically informed charting of bee disappearance, Carr's map chimes ominously with similar kinds of ecological diversity loss brought about by industrial agriculture. We might think again of Coffey's American bullfrog map where the bullfrogs escape their concrete pen with an exuberant flourish that indexes events in Latin America, Asia, and elsewhere. Indeed, the gestural force of the atlas was always intended to reach beyond the sum of its maps, towards incalculable numbers of similar moments and events—and the transplanetary effects that patches teach us to recognize. In this way, these single patches come to index the making of repetitive, characteristic patches through effects of infrastructure, such as industrial agriculture, or mining, or carbon-burning power stations. These accumulate and combine to shape the patchy assemblage that is the Anthropocene (more on this in Chapter 4).

Feral Atlas demonstrates how the practice of mapping—or countermapping—can provide vital analytic force. Encountered in no particular order within the digital site, these maps refuse a singular perspective and so perform the patchy Anthropocene argument in their disparate, overlapping, and heterogeneous depictions of the spatiotemporal dynamics of the Anthropocene, emerging from moments of spill, leakage, and feral proliferation. Charting co-creative, relationship-wrenching, natural histories in the making, they reveal the patchy Anthropocene in formation at varying speeds and scales.

A further exploration of time is long overdue. If we treat the Anthropocene as patchy, it becomes clear that different kinds of time are necessary to understand the dynamics of each patch. We're not just referring to separate timelines, such as varied starting dates for the Anthropocene or even the chronologies of various patches, but also to multiple modes of working with time. There may indeed be varied timelines, but these are not enough to see the patchy Anthropocene's temporal multiplicity.

Through what other kinds of temporality do Anthropocene patches come into being? Part II explores temporal ruptures as a feature of Anthropocene patchiness. Part III offers multistranded histories—and not just human ones—tangled into knots.

FIGURE 19. *Greenhouse.* El Ejido, Spain, 2013.
ARMIN LINKE

PART II
Ruptures

Part II turns to discontinuities in the time of the Anthropocene. The patches that make up the Anthropocene, we argue, involve time as well as space. Three kinds of ruptures are discussed in this section. Chapter 4 introduces ruptures in ecological and social systems. Chapter 5 turns to how individuals and communities experience Anthropocene ruptures—and particularly those of climate change. Chapter 6 discusses world historical ruptures.

The forms of temporal rupture introduced in this section rub against the temporality most commonly understood as defining the Anthropocene: that of geological epochs. Geological epochs encompass the whole earth. They have neatly defined beginnings and endings, which are key to their analytic use to classify geological strata. The beginning of a geological epoch is defined by a specified marker in earth's strata called a "golden spike." Like the spike that ceremonially marked the completion of the U.S. railway system, it forms a definitional point from which to assess the epoch.[1] The time of geological epochs is magisterial, ruling the planet as a whole. As geologist Jan Zalasiewicz puts it, "It is inherently synchronous within the domain across which it operates, which is that of the home planet."[2] The epoch sweeps exceptions under its cloak. After the golden spike, the epoch reigns completely until its end, fully covering the earth. In practice, geological periodization may have uneven traction across different regions; the Quaternary period, for example, is named after Ice Age gravels that appear only in the Northern Hemisphere. But the theory glosses over this to coordinate understandings for a single planetary history.

This is not the only temporality recognized in the practice of geology, which thrives on multiple kinds of time. The time conjured by fossil hunters, for example, is field-based, empirical, and full of gaps and false starts.[3] The time of mineral evolution is transformative, bringing new kinds of objects into the world through the contingencies of earth's history.[4] Yet, because epochal time defines the Anthropocene for geologists, some environmental humanists have endorsed the temporality of the geological epoch as the only one that matters.

Dipesh Chakrabarty's fidelity to the planetary scale, as a spatial unit, was introduced in Chapter 1.[5] His argument follows into time: Only planetary time, he argues, is appropriate to describe the Anthropocene; human timescales are fundamentally incommensurate with geological, planetary-scale time. Humanists and social scientists, he asserts, mistake the time of world history for geological time; they convert the "force" of planetary relations into "power" relations among human beings. Bringing capitalism or imperialism into studies of the Anthropocene only muddles planetary discussion. Geology drops out; moral affects and political prescriptions take over.

Chakrabarty's argument has some traction for the singular and homogeneous Anthropocenes called up by humanists and social scientists, such as sociologist Jason Moore's Capitalocene.[6] This is because it is difficult to chart the *geological effects* of global capitalist practices within a homogeneous planetary scale. If we move to the scale of patches however, the same problem does not arise. Consider mining, especially the recent form of mining in which a small amount of valuable material is sifted from an immense mass of overburden.[7] The geomorphological effects of mining are completely clear, not only in huge displacements of earth but also in the chemical changes that take place in tailings piles and ponds. The geological impact of mining is striking when we see the visual scale of overburden, as illustrated, for example, in a film by Anna Friz and Rodrigo Ríos Zunino showing the overburden of mining in northern Chile, the size of which rivals nearby mountains.[8]

The Anthropocene patches that emerge through mining are geological units of analysis. They are ecologically descriptive, rather than an imposition of the political. Just how they relate to the planet is another step in the analysis. We know that the earth moved by industrial practices exceeds that moved by glaciers.[9] But the key scale for analysis is not the planet but rather the patch. Mining creates significant patches within the planetary assemblage of the Anthropocene. Geology need not drop out. Nor do patch temporalities displace geological epoch making; they are a complement, allowing more detailed, closer-to-the-ground kinds of study.

Many kinds of temporalities combine in creating geological patches. In the patches formed by mining, for example, material lying deep in the earth, formed over time periods vastly outstripping human historical time, is extracted in a relatively short period of years, for purposes of infrastructural use and capitalist accumulation. Meanwhile, the discarded material and chemicals used for, and released by, these processes initiate new social-ecological disruptions. Such "knots" of multiple time are explored throughout this part and the next.[10]

FOUR

Hotspots in the Patchy Anthropocene

Something is broken—and life as we knew it is no longer possible. This is what we mean by "rupture," and we use the concept to consider how imperial and industrial infrastructures, repetitively built, have transformed ecosystems across the surface of the earth—and often beyond easy fixing. The place-time knots where "something is broken" are our hotspots. Although the word "hotspot" emphasizes place, this book argues that you cannot consider place in the Anthropocene without attuning to more-than-human histories. In this chapter we identify Anthropocene hotspots through the occurrence of rupture, that is, a mode of temporality in which the structure of emplaced relations itself changes. Previous rhythms disappear, and an entirely different temporal frame becomes necessary. The Anthropocene is full of ruptures; this chapter offers some tools for noticing them—and understanding the processes through which they occur.

It's always good to start with an example; here's a humble one, made clearer by a photograph. Pierre du Plessis describes the formation of bare-earth "sacrifice zones" with the coming of drilled boreholes to the Kalahari Desert of Botswana.[1] Boreholes were introduced to allow cattle raising to replace hunting; developers imagined boreholes as replacing

savagery with civilization. Meanwhile, the boreholes disadvantage the formerly hunted animals, that is, the wildlife that once thrived through migrating within the scattered and shifting water resources of the desert. Boreholes concentrate water resources for the maximization of cattle, which, in turn, eat the available grass, leaving sacrifice zones without the biological assemblage centered around grass. Grass, du Plessis documents, is a key resource not just for wildlife but also for ants and termites, underground truffles, soil bacteria, and much more.[2] Without this assemblage, the ecology changes.

Du Plessis uses a photograph (Figure 20) to explain the effect of the concentration of cattle on the desert ecosystem—in even a very short period of time. A barbed wire fence separates a wildlife management area, which cattle have trouble entering, from the area grazed by cattle. On the right side of the photograph, where the cattle graze, the ground is denuded. There is no grass evident, only leafless woody shrubs. On the left side, where the fence protects the vegetation from cattle and where only migrating wildlife graze, the grass—and thus all the insects, fungi, and other beings it supports—is healthy for this desert area.

FIGURE 20. **A divided landscape: Concentrating cattle around boreholes destroys graze for wildlife, as shown on the right side of the fence.**
PIERRE DU PLESSIS

Cattle gather around the boreholes, eating down the grass in the area directly surrounding the water. As boreholes proliferate across the desert landscape, wildlife habitat becomes scarce and fragmented, and more and more of the landscape is sacrificed to cattle. Boreholes also effectively privatize what was communal land, changing the social system. Both ecology and society are shifted by something as simple as the drilling of wells. The Kalahari's Indigenous San foragers are forced by these transitions (as well as new laws) to either adopt cattle or go hungry, with the wildlife. As Julie Livingston documents, this is part of a larger transformation of Botswana through water management—for commercial cattle production, or, as Livingston puts it, for "self-devouring growth."[3]

Here's another example involving water—but, this time, its loss. Michael Vine has written about Southern California's Owens Lake, which a hundred years ago hosted steamboats and irrigated fields and orchards.[4] Then the lake was piped into the water system of Los Angeles. As Figure 21 shows, Owens Lake became a mechanical element in a great system of water management, a system that stole water from across the region for the benefit of the city. There is nothing left of the original waterscape except sources and sinks for water managers. By the 1980s, great dust storms blew where the lake once lay. The dust seeped into residents' houses, even when they carefully sealed windows and doors. Asthma and other respiratory illnesses resulted. A lake setting became a dust bowl: This is rupture. Dust became a feral collaborator of the water infrastructure, creating the kind of anthropogenic disease environment that has come to characterize the Anthropocene.

The management of water has been a key feature of European expansion across the globe, and it has changed the nature of arid landscapes. Today, as aquifers are pumped more aggressively, the land sinks rapidly and water becomes scarce. Vine's report is one small but vivid episode of this history.[5]

Something is broken—and life as we knew it is no longer possible. Since the development of European empires, water management has been a project of governance, involving infrastructure that transforms

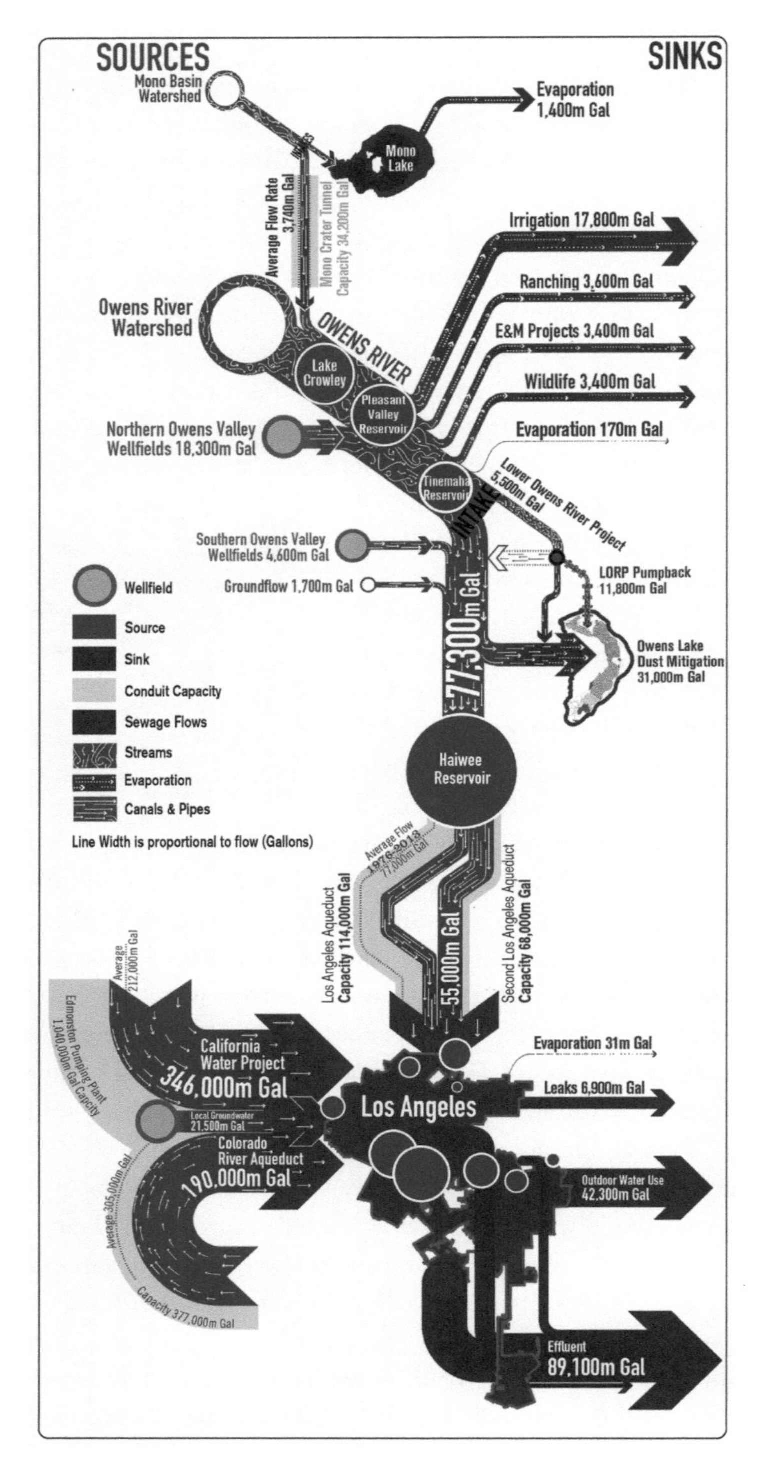

FIGURE 21. **This map visualizes the average water flows through the Los Angeles Aqueducts from 2003 to 2013.**
BARRY LEHRMAN AND THE AQUEDUCT FUTURES PROJECT

the land.[6] In Botswana and in California, water management has been important in the dispossession of Indigenous peoples.[7] As Vine reports, it is not a coincidence that Japanese were incarcerated in the dust storms of Owens Lake during World War II. There is reason, we think, to tell the stories of Botswana's sacrifice zones and Owens Lake dust ecology side by side. Each shows us a hotspot, a site of rupture, carelessly set in motion by regimes of state and imperial water management. One way to understand these ruptures is as a form of "systems" change (or "state" change). We turn to that analytic tool.

Systems Change Remakes Land- and Waterscapes

Systems are sets of relations that hold together, at least well enough to think with. Systems are experiments in thinking through relations. They help us understand relations among concrete entities, but the systems themselves are not concrete material beings. They are experimental holisms, helpful in understanding what's going on.[8] When we speak of "capitalism" or "families," we are positing somewhat coherent sets of relations, that is, systems. Not all systems are grounded in place, although some are, and these are particularly interesting for studying social and ecological assemblages. When systems change, their constituent relations undergo radical reorganizations and sometimes complete breakdowns. Everything we once thought we could depend on falls down a hole, from which it is difficult to climb out.

Because of the power of thinking about ecological relations within systems, it behooves us to think about environmental crisis in a language of systems ruptures—or, here, hotspots. To get started on the importance of systems rupture, one might consider an established patch (although patches are often made in ruptures, rather than preexisting them). Let's go back to the Black Sea in the late 1980s, discussed in Chapter 2. When comb jellies were introduced in the midst of overfishing, eutrophication, and bad weather, they took over. The ecosystem that featured fish disappeared; the baby fish were eaten by comb jellies before they had the chance to mature and reproduce. The system had changed, and the fish were no longer players. One might use the language of systems theorists to say that an "alternative stable state" had been established. This was the jelly world in which fish were only jelly food.

Systems rupture is a mode of temporality that differs from chronology or periodicity. It requires thinking in terms of a tentatively stable set of relations—and then the often-abrupt change to a different set of relations. In this same spirit as the Black Sea example, we can notice how European feudalism gave way to capitalism as an alternative stable state. This is a temporal phenomenon, but not a timeline.

Scholars in the environmental humanities are often suspicious of discussion of systems, because we associate it with abstract models in which often-doubtful assumptions are covered up by incessant quantification. But not all systems thinking involves abstract quantified models. Humanists use the concept of systems all the time, even if we don't admit to it. For example, we just mentioned the transition between feudalism to capitalism in the political economy of Europe. This is a system change. Marcel Mauss's differentiation between gifts and commodities also offers us systems;[9] if we posit a change from one to the other, we traffic in systems rupture. Ecosystems are of course systems. Whether or not we agree with a particular experiment in systems thinking, we can't help but find that kind of thinking everywhere around us. And, if one keeps an open mind, it might very well be a point where humanists and natural scientists find overlapping concerns. Changes in social and ecological systems can be tightly linked.

To see how systems change—that is, how rupture works—let's consider, once more, a delimited patch. Zachary Caple describes Lake Apopka in Florida, once one of the best fishing destinations for bass in the United States.[10]

> In March 1947, the first recorded algae bloom surfaced on the lake. Over a six-month period, the bloom consumed the whole lake. Previously clear water turned pea green. The algae bloom blocked light from reaching the eelgrass, dooming the macrophyte community. The death of the macrophytes then triggered an ecological chain reaction: when the eelgrass died, the bass disappeared (along with the fishermen) and the lake's sediments—formerly stabilized by the plant's roots—churned restlessly in the wind. This churning action resuspended phosphorus into the water column and nourished the growth of more algae. Lake Apopka had flipped into a new ecological state. Fish kills, mysterious wildlife deaths, and a foul odor plagued the lake for decades, earning it the notorious distinction of Florida's most polluted lake.

Lake Apopka became a different kind of lake: a cloudy, algae-filled one, instead of a clear one; a lake without bass and without eelgrass. Caple explains what happened: Fertilizer from surrounding agricultural fields drained into the lake, increasing phosphorus levels. For a long time, the lake was able to sustain its original ecological relations, involving bass and eelgrass; then, rather suddenly, the nature of the lake changed, and these lake denizens died. It's the comparatively sudden timing of this change that makes it useful to think of each lake form—clear versus cloudy—as a system that flips from one state to another. Each retains its relations comparatively stably. Once the lake has lost its clear character, it is not easy to remove enough phosphorus to bring it back.[11]

A lake changed by excess nutrients is probably the clearest, cleanest example of systems rupture; it harks back to the founding theory of alternative stable states in ecology.[12] Other systems are rarely so cleanly bounded. Still, watching systems change in patches, however open-ended and shifting, can lead us into larger units of analysis. The nutrients that changed Lake Apopka did not originate in the lake. Caple's research on the lake leads him to the phosphorus industry, from mining to usage as fertilizer, as this results, among other things, in the radical shifting of the lake. The phosphorus industry, in turn, comes into being within an agro-industrial complex, as discussed above. In studying the Anthropocene from the perspective of Lake Apopka, it is the complex of building projects that surround and suffocate the lake, such as mines and commercial fields, which come into focus. To understand systems ruptures, as these form Anthropocene hotspots, we need to look at the complex of infrastructure building that instigates particular spaces and times of disaster.

Imperial and Industrial Infrastructure Instigates Rupture

The infrastructures built by states and corporations in the last five hundred years have had exceptional force in disrupting ecosystems. How should we understand this force? Here we suggest that imperial and industrial infrastructures have the ability to multiply, that is, to inspire repetition across space as an integral feature of projects of conquest, governance, and capitalist expansion. To make this argument, we need to return to the definition of infrastructure that we rejected in Chapter

2: infrastructure as networks of makers and users, who together develop political communities and visions. While that definition obscures the out-of-network ("feral") effects of the infrastructure, as material form, it helps clarify why infrastructures have been so convincing, endorsed by both elites and ordinary people around the world. As long as infrastructures are understood as ways of getting important work done, they are difficult to question.

Infrastructures don't just proliferate because of their perceived benefits, of course; they respond to power dynamics built into state and corporate expansion. The standardization of infrastructure, which nullifies the particularities of local terrain, has been a common feature of imperial and industrial projects. Patrick McCully notes that only a few construction companies control the worldwide making of large dams.[13] Dutch engineers are famous for bringing their signature coastal infrastructure around the world.[14] Infrastructure-building projects extend far beyond state and regional boundaries through such transfers. Many industries, such as the oil industry studied by Hannah Appel in its instantiation in Equatorial Guinea, self-consciously adopt an ethos of "modularity," that is, the transfer of standard infrastructure packages from one place to another.[15] Standardization is promoted as a utopian form, that is, a form of progress. One way to understand the force of imperial and industrial infrastructures is in relation to their utopian claims, which disregard native communities and ecologies to entirely remake the terrain.

In some cases, colonial and postcolonial governments have modeled their infrastructure projects on those of earlier kingdoms. River management in the United States, for example, took instruction from British imperialism, which, in turn, learned from precolonial South Asian kingdoms.[16] In other cases, the goal has been to distinguish infrastructure projects from the past. Nuclear power, for example, has been sold as out-of-this-world futuristic governance.[17] In both cases, the goal of the most powerful infrastructure builders has been to develop exemplars that would be copied around the world. Imperial and industrial infrastructures are made to be copied. The highest dam, the most beautiful bridge, the most profitable irrigation system: Each of these begs to be repeated and perhaps surpassed.

The aspiration to build exemplars makes it worthwhile to bring the two meanings of infrastructure together: (1) infrastructure as material building project and (2) infrastructure as guide for visions.[18] The first definition allows us to see their effects: That's how they create ruptures. The second definition makes it clear that infrastructures are expected to do particular kinds of work: That's why they inspire. Here we need both. Classifying building projects by the kind of work they are supposed to do (their aspirations) allows us to watch their iterative proliferation—and the chains of rupture they set loose (their material effects).

Consider the agro-industrial complex, taken as a set of building projects involving not only fields but also the mining and production of chemical inputs, the machines, and the labor and property relations that service those fields. These support not just any fields; this complex creates fields designed for the maximum accumulation of wealth by investors, and this tends to mean exotic monocrops planted in machine-ready rows for economies of scale. Chapter 2 introduced the research of Ivette Perfecto, who studied the transformation of smallholder polyculture coffee planting in Central America to favor monocrop full-sun plantations. This is agriculture under the sway of the agro-industrial complex. As explained in Chapter 2, this change brought about a rupture: a systems change in which coffee rust fungus infected peasant farms and commercial plantations alike.

To see the force of the agro-industrial infrastructure complex, Perfecto's research is worth reading alongside other accounts from this same complex, including Alyssa Paredes's research on chemical cocktails sprayed on banana plantations in the Philippines, introduced in Chapter 1. The chemical cocktails spewing from the planes may be aimed at the banana fields, but they drift over surrounding homes and farms. The rupture here is in multispecies health: Local farmers report new forms of death and illness in their animals, their crops, and their children.

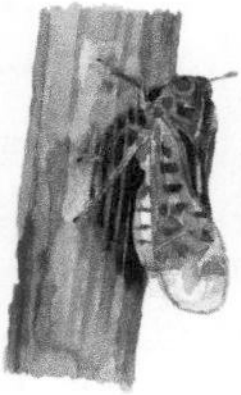

We might add to these reports Elaine Gan's study of the brown planthopper in the Philippines.[19] With the green revolution, chemical inputs in rice agriculture managed to turn a once relatively innocuous pest into a superpredator: a much larger and more adaptable insect, nurtured by a high-nitrogen diet, and capable of taking down all of a field's crops. Gan calls the transformation "world changing" and tracks how it emerged from the conjuncture of changes in plant breeding and fertilizer development. From the conjuncture, a rupture: Neither great commercial farms nor smallholders can grow rice today without the threat of this superpredator.

Using Tippers to Notice Ruptures

In *Feral Atlas,* we used the term "Tipper" to explain the power of imperial and industrial infrastructures to attract feral partners that stimulate social and ecological ruptures. With a nod to ongoing efforts to describe the Anthropocene as a series of ecological tipping points, we use Tippers to identify clusters of modern infrastructural work that instigate state changes. Rather than just dealing with infrastructure-mediated systems change one case at a time, in *Feral Atlas* we argue that there are benefits to working with clusters of cases, grouped in relation to the kinds of work infrastructures are designed to do. The *Feral Atlas* Tippers are BURN, CROWD, DUMP, GRID, PIPE, TAKE, and SMOOTH/SPEED, each intended as a verb, and each corresponding to aspirations toward a particular kind of work. PIPE, for example, groups projects of water management; TAKE groups systems change due to long-distance imperial and industrial transport; GRID groups effects of ecological simplification in agriculture and livestock rearing.

Such clusters are heuristic; they encourage *Feral Atlas* users to identify infrastructures around them as sources of ecological systems change. By grouping cases in relation to infrastructural work, *Feral Atlas* draws attention to the point emphasized above: Imperial and industrial infrastructures are built in repeating patterns, and in relation to power and desire.

To show how these infrastructure-building complexes gain force, *Feral Atlas* includes a series of short and stark video poems.[20] Each shows the infrastructure—without commentary and mainly without human presence. Each evokes the enormity of its effects merely by its simplicity. It's worth presenting a couple of these to show the usefulness of the Tipper concept.

Here, for example, is a still from one of the videos for BURN, shot by Isabelle Carbonell and edited by Duane Peterson (Figure 22). The video shows a gas flare—and nothing else. Imagine a world emptied of everything except this gas flare . . . and then refilled only by its feral effects. What have we allowed burning gas to do? The videos, like the Tippers themselves, are thought experiments in which we ask how infrastructures such as this gas flare remake the ecosystems around them, big and small, and from there, the world.

Consider another screenshot from a Tipper video, this one an exposition of GRID, that is, the infrastructure of the agro-industrial complex. Figure 23 is a still from a video poem shot for *Feral Atlas* by geographers Trevor Birkenholtz and Bruce Rhoads over a cornfield in Illinois, and edited by Armin Linke, who made many of the *Feral Atlas* Tipper videos. Birkenholtz and Rhoads also wrote an informative essay about the making of this grid, including the extensive drainage system that allowed cornfields of this sort to emerge; the agro-industrial complex completely remade the land.[21] The image offers nothing but the kind of field we have learned to naturalize—a monocrop field in straight rows created for the work of machines or alienated labor. Imagine a world in which there is nothing but these lines and lines of industrial corn. This world destroys the previous landscape and attracts the feral ecologies industrial corn creates. Look ahead to Chapter 9, which describes pathogenic fungi as Anthropocene collaborators, to conjure images of this field in 1969 when southern corn leaf blight infected corn across the U.S. Midwest. The corn—and thus the economy of the region—was

FIGURE 22. **Video still from *Natural Gas Extraction*. Qatar, 2008.**
ISABELLE CARBONELL AND DUANE PETERSON

FIGURE 23. **Video still from *Corn Harvest*. Champaign County, Illinois, 2018.**
BRUCE RHOADS, TREVOR BIRKENHOLTZ, AND ARMIN LINKE

destroyed, leaving only sweeping gray clouds of spores. It was something a science fiction movie might help us imagine. What have we allowed industrial corn to do? The video, which presents the corn without these feral collaborators, is yet haunted by feral possibility. It offers a thought experiment about patches, systems, and worlds. It displays the dangers of a world in which feral effects usurp the earth's ecologies, taking over, patch by patch.

The Fossil Fuel and Nuclear Power Infrastructure Complex

Identifying ruptures as resulting from the effects of particular infrastructure complexes, that is Tippers, allows some strange meditations. For example, *Feral Atlas* lumps together fossil-fuel generated power and nuclear power, despite the fact that they are most commonly presented as opposing each other. This opens reflection on what varied energy-production infrastructures have in common. The carelessness of one is, indeed, reflected perfectly in the other. In both cases, the energy needs imagined as fueling progress (here is infrastructure as aspiration) allows destructive building projects (here is infrastructure as material form) with devastating feral effects.

Consider two kinds of "overheating": the effects of excess carbon dioxide, caused by burning fossil fuels, and the effects of radioactive effluents, distributed from nuclear plants. Although they are technically quite different, there is surprising common ground. First, while both are patchy in their initial making, both reshape the whole planet. Carelessness in each industry is encouraged by human inability, in both cases, to corral feral effects. Second, both shape deep-time futures in which humans may no longer inhabit the earth. It is hard to think of any environmental effects quite like these in the extensiveness of their space and time ruptures. This alone makes it worthwhile to think of these two energy sectors together.

Geologist Jan Zalasiewicz, working with artist Anne-Sophie Milon, conjured specters of each of these two space-time effects in a report on carbon dioxide gone feral.[22] Patchy yet planetary: Burning fossil fuel is patchy; it is concentrated particularly in northern cities, where industry as well as transportation create great clots of carbon dioxide. Although it takes some time, these clots dissipate around the globe, creating a planetary condition of overheating. Hard to corral: Some parts of the earth, such as the Arctic and Antarctic zones, have been much more quickly affected by climate change than other parts, despite their distance from fossil-fuel burning centers. Milon and Zalasiewicz focus on the patchy distribution of the "victims" of anthropogenic carbon dioxide; for example, coral reefs have been hard hit by rising ocean temperatures. Coral, unable to retain its symbionts under hotter conditions, is bleaching across vast areas. Deep-time futures: "To arrive back at a preindustrial-style atmosphere by planetary mechanisms alone would take about 100,000 years."[23] Carbon dioxide can eventually turn to rock within the surface of the earth, but only after millennia.

Let's compare this example with the work of Norio Ishida and Daisuke Naito, who write about the radiation that has spread in and beyond Japan since the breakdown of the Fukushima nuclear power plant.[24] On March 11, 2011, an earthquake and then a tsunami hit the reactor complex, discharging radioactivity across a broad region, as shown in Figure 24. But this was hardly the range of contamination—even without counting radioactive material released or washed out to sea. The Japanese government compounded the problem of radioactive contamination by allowing materials from the Fukushima area—from vegetables grown in radioactive soil to wood chipped from trees killed by radiation—to be distributed across the country. In reaction, Ishida's lab began a citizen science project in which people could bring soil and other samples from their region for testing. They found that wood chips from Fukushima had been laid

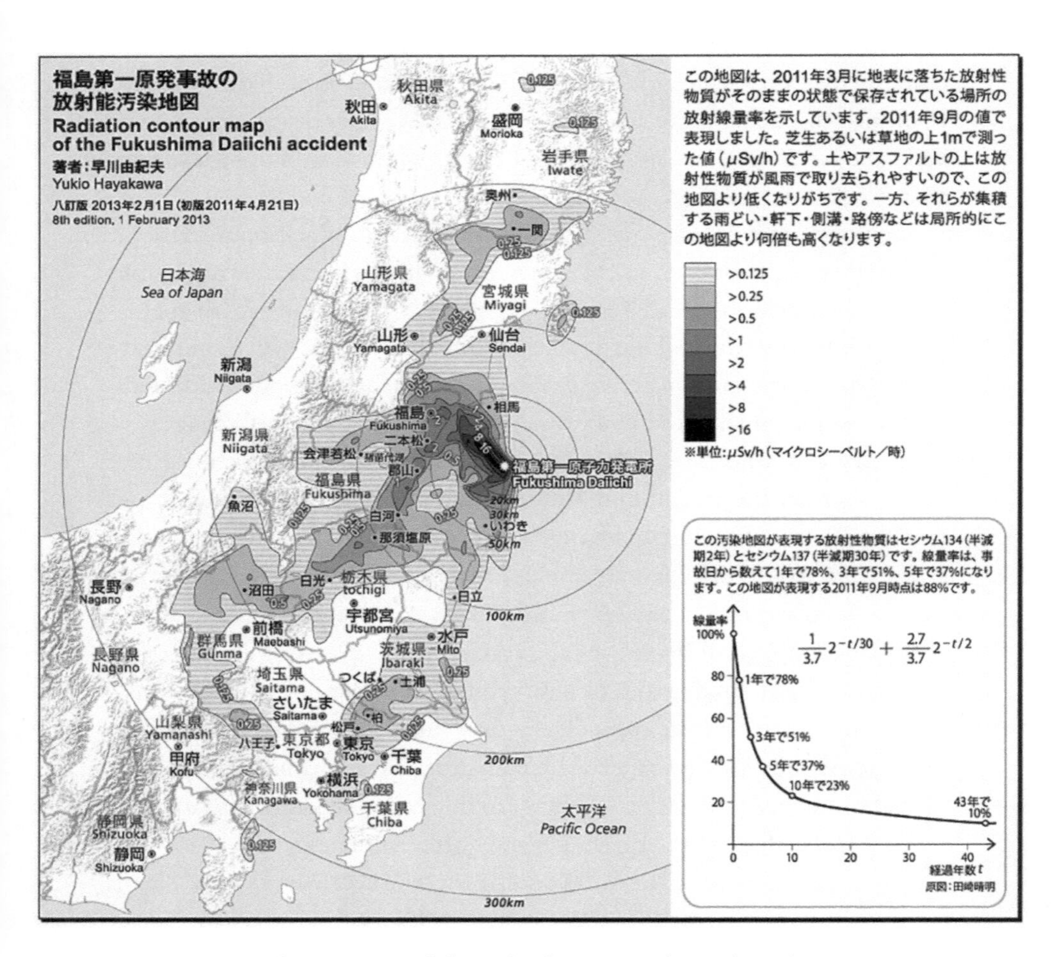

FIGURE 24. **Radiation map of the Fukushima Daiichi nuclear disaster.**
YUKIO HAYAKAWA

down as part of a streamside restoration in the area near Kyoto. The developers had measured the radioactivity when the chips were saturated with water, pushing it under the limit; when dried, the chips were much too radioactive for legal use. A suit ensued, drawing the case to public attention. Further research followed, including the discovery of releases of plutonium 239 and 240 from the Fukushima plant. Plutonium 239 has a half-life of 240,000 years.

Again, the same characteristics. Patchy yet planetary: The radioactivity is generated in a particular patch, but it spreads widely to create other patches, indeed encompassing the planet. When Finland reported Fukushima radiation, this required measuring radiation spread by winds that had made it all the way around the globe. Hard to corral: The "victims" are heterogeneous across types of matter and across space. Some organisms, including many plants and fungi, take up radioactive cesium as part of their metabolism. Distribute those plants and fungi, and these effects have been spread. We humans too absorb it; our bodies become saturated with it. We carry its system effects. Deep-time futures: The spread of radioactivity is changing the chemistry of the earth for a very long time to come. It will surely outlive our species.

The infrastructure built out of the search for industrial energy (as well as for war) creates system ruptures with catastrophic effects across the planet. By considering these examples together, it becomes possible to contemplate the vast scope and scale of systems rupture caused by industrial and imperial energy production.

The DUMP Complex

Considering waste management (and non-management) as a complex of infrastructure also offers insights; this is the work *Feral Atlas* attempted in suggesting the Tipper DUMP. The cases considered together under the Tipper DUMP show us how waste has changed: Whereas most human wastes in the past were biodegradable, today many synthetic products can last for millions of years. Furthermore, new kinds of toxins have been introduced. In the past, toxicologists tended to assume that the poisonous quality of a substance was directly related to its dosage in the body; now whole classes of substances, such as hormone disruptors, are effective poisons at any concentration. In addition, many of the new long-lasting toxins accumulate in bodies, so that doses in the past are added on to current doses. One further characteristic of Anthropocene waste explored in the DUMP complex: These toxins—set into motion by humans—are then redistributed by nonhumans, both living and nonliving. This set of problems makes it useful to look at quite heterogeneous waste materials and processes side by side.

Kelsi Nagy considers how discarded plastic bags enter human consumption in the city of Mysore, Karnataka, in southern India.[25] Urban dairy cows wander around the city; they are valued and often hand fed. But, from the cows' perspectives, the most delectable meals are those humans leave behind, and these are often in plastic. (Nagy found that the cows prefer the quick calories of food from dumpsters to fresh grass.) Plastic accumulates in the body of the cows. Much of it does not disintegrate at all, but it sheds phthalates, the plasticizers that make polymer chains supple and that are well-known hormone disruptors. Phthalates are also found on the grass cows eat around plastic. Meanwhile, they—and other toxic chemicals—are secreted in the milk of these cows, which is considered especially healthy by consumers, because it is local. Milk contains liquid essences of plastic. This is waste in the Anthropocene.

In a related meditation, poet Evelyn Reilly considers the nature of Styrofoam. Here is an excerpt:[26]

> Enter: 8,9,13,14,17-ethynyl-13-methyl-
> 7,8,9,11,12,14,15,16-octahydro-cyclopenta-diol
>
> (aka environmental sources of hormonal activity
> (side effects include tenderness, dizziness
>
> and aberrations of the vision
> *(please just pass the passout juice now!)*
>
> Answer: It is a misconception that materials
> biodegrade in a meaningful timeframe

Answer: Thought to be composters landfills
are actually vast mummifiers
of waste
and waste's companions

Thinking of a dairy cow and a landfill together turns out to have some usefulness: Each allows us to consider the particular ways that infrastructures of waste change the possibilities of life in the Anthropocene. This is the work of Tippers classifications in *Feral Atlas*: By attending to complexes of infrastructure projects, it is possible to productively connect seemingly disparate systems effects. We hope readers will play further with this digital feature of the atlas, and perhaps even read the explanations (e.g., the extended explanation of "state change") we offer there. For all of us, the challenge remains: If we want to better understand how the Anthropocene has been so effective at disrupting social and ecological systems, we would do well to think more about proliferating, repetitive complexes of infrastructure building.

Systems Ruptures and Anthropocene Hotspots

Systems ruptures create Anthropocene hotspots, that is, zones in which previous social and ecological relations no longer hold. Many kinds of evolutionary relationships have developed over millions of years, allowing organisms of different species to live together. Systems changes that threaten these evolutionary relations change the habitability of the place not just for one species but for many. Many plants, for example, need insect pollinators. Kill off the pollinators, and the plants can no longer reproduce, as Marcela Cely-Santos's research illustrated in Chapter 3. It is not easy to come back from system changes. As Ivette Perfecto discusses for the plague of coffee rust fungus in Central America, once the rust spores have saturated the area, polyculture no longer suffices to protect coffee trees. Just planting a few more trees will not end the epidemic. As Zachary Caple shows, just removing a little phosphorus from Lake Apopka will not return the habitat for bass. This is a general problem with systems ruptures: They change relations enough that they are hard to undo. These are dangerous zones for many kinds of life.

Some natural scientists have tried to identify tipping points at the planetary scale.[27] This is a disputed exercise precisely because of the heterogeneity of the ecological systems that make up the planet.[28] However, efforts to find planetary tipping points show how even those completely dedicated to the planetary scale are also able to explore alternate modes of temporality, including the time of systems rupture, rather than merely staying inside the magisterial synchronicity of the geological epoch. This section goes on to explore two other kinds of systems-oriented temporalities: the rupture of seasonal expectations, in Chapter 5, and the rupture of world systems, in Chapter 6.

FIVE

Unseasonal Weather

ALDER KELEMAN SAXENA

How does rupture feel?

It is April of 2020, and I am fighting panic. I am in Flagstaff, Arizona, dutifully staying at home with my family during the early days of the COVID-19 pandemic. We are far enough into the crisis that Arizona's governor has issued a stay-at-home order, but it is also early enough that it is clear the worst is yet to come. The peak of the epidemic curve. The threat of a second wave. The economic aftershocks. I have two small children, and my academic employment is precarious. My family is far away. My husband's family is farther. We worry about who would care for our kids, were we to get sick. We worry, too, that our family members might become ill, too distant for us to help.

It will get better in August, I think. And then I think about fire.

In 2019, the year preceding the pandemic, Flagstaff's forest fires were intense. Not as intense as California's but concerning enough to result in evacuations of neighborhoods where backyards border Forest Service land. Concerning enough to spark fears about flash floods during the summer monsoon—a sudden rush of water scraping exposed topsoil and debris and channeling them down one of the seasonal waterways that run through the center of town. Last year, a small child drowned in one such flood, a few blocks from my son's preschool. Now, when there are flash flood warnings, they close school early.

Flagstaff is nestled in the world's largest Ponderosa pine forest, a fire-mediated ecology. But like elsewhere in the American West, it is clear here that the shift to what Stephen Pyne calls the "Pyrocene" is bringing new frequency, intensity, and risk to forest fires.[1] This is not the predictable, renewing fire of times past.

In these early days of April, I let my mind dart, fearfully, across scenarios: the connectivity of tree canopies across neighborhoods; the few highways that lead out of Flagstaff to other places; pressures on institutional capacities across sectors, and where those sectors connect.

Hospitals, paramedics, firefighters. Fire, flood, plague.

Don't get ahead of yourself. I take a mental breath. None of this is a forgone conclusion. We might flatten the curve. It might be a cool, wet summer. I might spend June and July teaching my children about the wildflowers that burst to life in fire scars. *Things will get better in August.*

But will they?

Toward a Material Phenomenology of Patchy Climate Change

Though the Anthropocene is not a product of climate change alone, thinking with climate change provides a particularly useful entry point for demonstrating how a patchy place-time analysis can intersect with the analytical strengths of social science and humanism.[2] Climate change, in the 2020s, is more than a concatenation of numbers collected from the world's thermometers, or a series of troubling graphs indicating statistical probabilities of abstract future events. Two decades into the twenty-first century, climate change is a lived experience: one that merges abstract knowledge with the sensory and emotional power of ecological disruptions, like unseasonal weather. This is a planetary phenomenon of which we all have situated knowledge.

But situated knowledge of a global phenomenon is a tricky thing. Climate change, by definition, is collective and far-reaching. It is an emergent property of actions that stretch across space and over time. It results from the activities of many different actors (think: commuters, parents, CEOs, politicians), and it manifests itself in a wide variety of ways (think: hot summers, glacial melting, extreme weather events).

Attributing these manifestations to climate change has always been slippery business. Because of the way that climate science is built—on large-scale, multisite, multiyear data sets, identifying aggregate trends over time—scientists have had a notoriously difficult time answering yes to the question, "Was this weather event caused by climate change?" Instead, following the conventions of careful research, they say some version of "Yes, probably; this particular event is the kind of thing that is likely to become more common under changing climatic conditions." Asking a climate scientist whether a particular storm is "caused" by climate change is a bit like asking whether the setting of the sun "causes" crickets to chirp in the night. While the events stand in close association to each other, they are interlinked by complex sets of relations not visible to the naked eye, a multidirectional tangle of influences. The designation of climate change as a "cause" of localized environmental shifts, while useful for politics and policy, talks at cross-purposes with the careful language of probabilities, likelihoods, and correlations that have defined scientific debates about climate.

It is difficult for situated knowledge to gain a toehold on this intellectual terrain. Without the backing of numbers (like rainfall statistics) or historic time-series of temperature and humidity, situated experiences of changing climate are easily dismissed as anecdotal: interesting illustrations, but subjective and untrustworthy. The perspective of climate change as something "dwelled within" can never, by definition, encompass the planetary process.[3]

This disconnect has by no means curtailed social science research on climate change, but it has constrained it to more comfortable terrain. Social scientists' approaches to the study of climate cluster around two major tendencies. The first is extremely grounded, documenting lived experiences of climate change in specific environmental contexts. For example, detailed descriptions of the impacts of climate change on Indigenous or other remote communities, which span the academic and practitioner literatures, exemplify this approach.[4] The second major tendency, often building on studies of lived experience, rests on a critique of knowledge-power relationships that emerge from the climate sphere, pointing out that "the global" is socially constructed, and constructed in ways that benefit particular groups to the detriment of others.[5] This approach positions social analysis of climate change within larger histories of social and economic inequality and contributes to the growing advocacy for critical climate justice.[6]

In this book, we advocate for a third way of looking at climate: one that sees climate change as *constituted* through its local effects. The emergence of climate change—as an assemblage of material and ideas—is about more than just local responses to external forces. It is also about how those forces interact with things-already-in-place (consider: organisms, ecological relationships, livelihood practices, institutions of power, and social meanings). This approach can both incorporate locally grounded analyses of the material effects of changing climate *and* critique global knowledge-power complexes; but it adds to these by emphasizing that the on-the-ground effects of climate change are not merely manifestations of uniform planetary forces. The fact that these forces manifest differently—responding to particular material and ideational conditions and layering into particular sets of relations—is a fundamental aspect of what climate change *is* in the current historical conjuncture.

Put another way, *to center situated knowledge is to understand climate change itself as a patchy phenomenon.* That is, the material manifestations of climate change are unevenly distributed, and differently realized, across landscapes; and this uneven distribution is not merely a challenge to be overcome by synthetic analysis, but rather a constitutive characteristic.[7] One might think of the materiality of climate change not just as part of a planetary whole, but also as a variegated fabric with different patternings in different places. Living amidst these patterns feels different from site to site—for example, extremely hot summers are a different lived experience than one hundred–year floods—and these particularities elicit new patterns of action by humans and nonhumans alike. Extreme temperatures drive shifts in cropping patterns—and in the dynamics of agricultural pest populations. Soil responds to flooding in ways that depend, in part, on soil type, and also on land-use history and water infrastructure. People make sense of—and respond to—these changes in ways that involve not just other species and nonhuman agents, but also their families, their faith, their governments, and the financial sector. The patternings of climate bring nonhuman beings into relationship with patterns of human thought, behavior, institutions, and social relations.[8]

While the agentive forces of nonhuman actors are the subject of most of this book, this essay focuses on what living amidst this patchy patterning feels like from a situated, human perspective. From a first-person point of view, experiences of the material and the ideational occur simultaneously, layering into each other. Taking these situated perspectives on climate seriously requires approaching them through what one might call a *material phenomenology*, or thinking simultaneously with the lived experience of the biophysical world and the worlds of discourse that we access through, for example, ideas, conversations, and social media.[9]

Drawing on my own field research in Mexico and Bolivia, and lived experience in Arizona, this essay considers what it might look like to think with a material phenomenology of climate change—or a "view from a patch."[10] It builds on the previous chapter in that it considers the manifestations of climate change to be a particular kind of rupture: a rupture in our experience of materiality-in-time. Climate change offers especially relatable examples of shifting material patterns within a temporal framework, in that it changes the timing and intensity of temperature, precipitation, humidity, and extreme weather events. Put differently, it makes weather unseasonal. These shifts change not only our material, lived experiences of such events, but also human *expectations* of what the future will be like. This chapter gives examples of the situated experience of unseasonal weather in three contexts and ends by returning to patchiness—considering how this approach draws attention to the convergence of the material and the ideational in useful ways.

Rain, Irrigation, and the Drug War in Sonora, Mexico

"No son así las lluvias!" *The rainy season is not supposed to be like this!*

Nearly two decades have passed, but I still remember the indignation in Doña Leticia's tone as she looked up at the clear, blue morning sky in Álamos, a small town in southern Sonora, Mexico. Though spoken in a casual moment, while hanging laundry on the internal patio of her house, where I was renting a room, the assertion was anything but off-handed. Doña Leticia was a formidable woman, accustomed to seeing her words spur action. In late July, at the start of the rains, the sky should not be blue, she huffed, as if the force of her will would urge the elements to correct their behavior. In the morning it should be overcast. The sun should not come until later in the afternoon.

The summer I rented a room from Doña Leticia was my first experience studying small-scale farming. As with many first encounters in fieldwork, much has become clearer to me in hindsight that I did not understand at the time. It was the early 2000s, and I was researching dryland maize farming around Álamos, with the intent of assessing the extent to which local farmers were continuing to plant landrace varieties of maize.

It did not take long to realize that not many people in my field site were interested in the idiosyncrasies of maize biodiversity. Farmers I spoke to were kind enough to humor me, but over time it became apparent that they saw maize cultivation as an ancillary practice—one that supplemented their main cash and fodder crops (sesame and sorghum) but that seldom satisfied the needs of the household, let alone bringing in any serious cash. Farming was hard in this region, and households practicing rainfed agriculture were grindingly poor.

And yet, somehow, some people still seemed to have food, to drive big pickup trucks, to drink beer, to throw parties. They were earning income from somewhere. As I became more attuned to the conversations taking place in subtext—through double meanings and clipped phrases—I learned to read another form of economic activity shoring up livelihoods, one based on the cultivation, transport, and trade of marijuana, poppies, and cocaine.[11] As it would turn out, I was hearing these murmurings in the early days of Mexico's drug war, during the fraying of longstanding alliances between politicians and cartels under the Vicente Fox government. A few years later, these disagreements would break into an all-out armed conflict under the government of Felipe Calderón.

At the time, if I had been asked about the connections between climate, farming, and the drug trade, I would likely have shrugged. Then, perhaps, I would have remembered the pickup trucks loaded with spools of irrigation tubing, which I sometimes saw bumping along roughly graded roads. They were headed, I was led to understand, up into the mountains, to irrigate the real cash crops. Meanwhile, farmers growing maize, sorghum, or sesame seldom had access to any water other than the summer rains. While one form of agriculture was benefiting from significant private investment (and offering high wages to those willing to work the fields), engaging in the other entailed climatic risk and economic struggle.

Now, many years later, this small detail seems far more salient, an infrastructural hint at a larger web of relations. Those irrigation lines—and their absence—tell a story linking the precarity of dryland farmers to the long-term neglect of local and national governments, which saw small-scale agriculturalists as backwards and needing to be absorbed into the urban economy. Álamos is located in the Mayo River watershed, just upstream from a major dam, the Presa Mocúzari, constructed in the 1950s. Irrigation provided by the dam was one of the factors that drove the development of the green revolution in the nearby Yaqui Valley on Sonora's coastal plains. But while the abundance of irrigation had touched off an agricultural boom in the Sonoran lowlands, it made dryland farming, by comparison, seem an anachronistic practice, and by the mid-2000s it had little support or acknowledgment from the state or national agricultural sectors.[12] It was under these circumstances, some forty years later, that the parallel state structure of the drug trade found footing. Where government bureaucracy and self-reliance failed, the cartels were willing to step in, offering not just jobs, but also social safety nets. They did so in exchange for loyalty, and violations of this loyalty were violently and publicly enforced.

Recent reporting in *The New Yorker*, detailing the links between climate change and the crisis of U.S.-bound migration from Central America, vividly traces the kinds of relationships I observed in Sonora so many years ago.[13] It points out that the notion that climate is a direct driver of migration is overly simplistic. Rather, localized crop failures make families more vulnerable, spurring them to migrate in order to secure income, and in turn requiring them to take loans to cover the costs of transport to the United States or elsewhere. Often, this financing comes from usurious lenders, putting them in debt, financially and socially, to actors in the illicit economy. This raises the stakes of the endeavor; detention, deportation, or any other form of failure risks the life and livelihood not just of the individual migrating, but also their friends and family at home.

In such settings, shifting climatic conditions exacerbate vulnerability, layering onto other forms of precarity. Just as ruptures in the viability of rainfed farming caused by climate change shift farmers' relationships with their more-than-human surroundings, they also shift their human relationships: with friends and neighbors, and with larger institutions, both formal and informal.

Viewed through this lens, many dimensions of the human tragedies of twenty-first century migration—separations, betrayals, untimely deaths—are also stories of rising temperatures, sparse rainfalls, and extreme weather events. To understand these stories—to think with a material phenomenology—ethnographers might start with situated knowledges of rainfall patterns, crop varieties, and landscape histories. But rather than stopping there, they might then ask how these experiences of climate change intersect with other spheres of lived experience. Where and how does climate emerge in popular culture, like the lyrics of *norteña* ballads, or Spanish hip-hop? How is the pope's encyclical on climate change being interpreted in the sermons of local Catholic churches? Or, how does farmers' trust in institutions—government or otherwise—inform their assessment of the relative benefits of experimenting with climate-resilient agricultural techniques promoted by local NGOs, versus choosing to migrate? Thinking with these kinds of questions helps to understand the contours of climate change from a patchy perspective, tracing what one might call an "emotional political ecology" of the lived experience of climate change.[14]

Climate, Potatoes, and Politics in the Bolivian Andes

Thinking with these layered spheres of experience helps to show how differences emerge within the global phenomenon of climate change. For small-scale farmers in Bolivia, as in dryland farming in northern Mexico, the relationships linking climate to agrobiodiversity are salient, but the social contours of these processes are different. Whereas the northern Mexican farmers with whom I did research so many years ago expressed only weary interest in maize, the Bolivian farmers with whom I worked a decade later enacted the "interspecies affinity" between humans and crops differently.[15] Bolivian farmers took pride in agrobiodiversity—not just of potatoes but also other Andean tubers and altitude-adapted, or "Andeanized," varieties of species originally introduced from Europe. They proudly displayed these varieties in markets and seed fairs. Some of the farmers I knew—who were quiet, at times hesitant, and given more to action than to words—would break into huge, unreserved smiles when encountering a beautiful potato.

By the early 2010s, though, climate change was having a marked negative effect on potato agrobiodiversity in the region of Cochabamba where I worked. Warming temperatures were opening higher altitude terrains to agricultural pests and diseases, like potato blight (*tizón*); and the increased pest and disease incidence was having a particularly negative effect on the minor, farmer-improved varieties of potato known locally as *papa nativa* (native potatoes). Many other, more widely planted varieties showed greater resistance to pests and diseases. In some cases, this resistance was due to farmer selection, and in others it was because the variety had undergone formal improvement by Bolivian plant breeders in preceding decades. This was a curious and troubling reversal of fortunes, particularly given that one of the scientific justifications for conserving agrobiodiversity has been the idea that diverse genes represent a reservoir of resistance against pests and diseases.[16]

But here, changing patterns of temperature and precipitation are not the only story. The pests and diseases potentiated by climate change had the groundwork laid by longer processes of agricultural modernization. My area of fieldwork, Colomi, sits on the major highway connecting the Bolivian Altiplano (and the capital city of La Paz) to the Amazonian lowlands of the country and to Brazil. As such, it was a major trade route for potatoes, and for everything else. Farmers in the area connected the opening of the road, in the mid-1970s, to the increasing commercialization of potato farming; local households produced potatoes not only for subsistence, but also for distribution to the regional and national potato markets.

This commercialization led to an increased intensity in land use, and a concomitant reduction in fallowing. In other words, it reduced the fallowed spaces between cultivated parcels, and the fallowing times between rotational uses of a single piece of land. The acceleration of cultivation effectively undercut the historic tools of pest management; it allowed pests and diseases whose spread might have been isolated by landscape barriers to easily hop from plot to plot, or season to season. This, combined with shifts in timing of planting (a response to the shifting reliability of rainfall), made the young potato sprouts particularly vulnerable to infections. In Colomi, while some pest-resistant native potato varieties were undergoing a boom in commercial demand, farmers were deciding, pragmatically, to simply not waste their time or effort on the rarest and least pest-resistant varieties. In this way, potato agrobiodiversity was threatened not just by something one might construe narrowly as "climate" (that is, by changing patterns and timing of rainfall), but also by changes in a constellation of other relations, including pest populations, commercial trade in potatoes, and patterns of land use within families and communities.

To the extent that agrobiodiversity was an object of admiration in rural Bolivia in this period, the pride farmers expressed for it fit into narratives of resistance to the culture of dominant elites, and of survival against all odds, not unlike the narratives that develop around narcotrafficking in Mexico. But in Bolivia, since the early 2000s, this narrative groove has taken a different shape, interweaving with the story of Indigenous resurgence under the leadership of former President Evo Morales. Morales became known as a firebrand in international spheres, not least for his pointed interventions in climate negotiations. In these debates, Morales staked out a position that advocated taking carbon-emitting nations to task, decrying the climatic disruptions already being felt by high-altitude farmers in this landlocked nation-state and demanding that those responsible for the majority of historical emissions (that is, the United States and European nations) shoulder the burden of reducing carbon output, rather than shifting this burden to developing countries. His advocacy abroad reflected a key discursive thread of his politics at home, which centered around the idea of re-founding Bolivia as a rightfully Indigenous-led state.

Morales's record of concrete policies supporting biodiversity in Andean agriculture was debatable, but his actions on the national and international stages opened up a narrative in which the practice of Andean agriculture, founded on the cultivation of the potato, could be seen as part of a larger history of Indigenous resistance. During Morales's tenure, rural agricultural activities could be recast as a form of resistance to five hundred years of external colonialism by the Euro-American West, manifested most recently in the shape of changing climates. Under these circumstances, while climate change created material vulnerabilities, Andean farmers' experience of those vulnerabilities layered into other forms of sociopolitical meaning making, most notably the populist wave that swept Morales to power in 2005 and maintained his presidency for some fifteen years.

In the aftermath of Morales's ouster in late 2019, the future of Bolivian politics is unclear, but the narratives around climate, agriculture, and Indigenous resistance fostered by his presidency seem likely to endure. These convergences between farmers' situated experiences of climate change and their lived experiences of the politics of racial exclusion in Bolivia are another formation of a material phenomenology, and one that may help in interpreting new developments in Bolivian politics—and agriculture—as Andean weather patterns continue to shift.

Extreme Weather in Northern Arizona

Fire, flood, plague. When I wrote these words in early 2020, it had the feeling of catastrophizing. As it turns out, I wasn't so far off.

As everywhere else, the COVID-19 pandemic did put pressure on the health care system in Northern Arizona. The impact was particularly acute on the Navajo Nation, to which Flagstaff is a border town. In the news media, the rapid transmission of COVID-19 among Navajo (Diné) people was attributed to attendance at charismatic gatherings, which drew relatives from larger cities.[17] Around Flagstaff, a small city with a large Indigenous population, people more commonly link the spread to the limited food distribution infrastructure of the Navajo Nation. There are only 13 grocery stores in the entire 27,000-square-mile territory, and my friends and colleagues identify these necessary gathering sites as likely points of disease transmission during the pandemic.[18]

As elsewhere, the effects of COVID-19 on the Navajo Nation layered onto preexisting health disparities, exacerbating the pandemic to a degree that famously led the nation of Ireland to donate international aid.[19] At the time of writing, in March of 2023, out of a resident population of around 173,000,[20] the Navajo Health Department reported over 82,000 cases of COVID-19 and 2,072 deaths attributed to the virus.[21] Estimates from Navajo County suggest that mortality rates from COVID-19 in 2020, before vaccines became available, were more than six times higher among Native Americans diagnosed with the illness than among white patients.[22] As *High Country News* reported, information gaps between tribal and state health care systems may mean that the full toll on Diné people (or other Indigenous Americans) will never be known.[23]

So there was a plague; and there has also been fire. In April of 2022—well before fire season typically starts—the Tunnel Fire burned across more than 21,000 acres northeast of Flagstaff, destroying more than 30 homes.[24] The fire started on a weekend, in the Kachina Peaks Wilderness. On a Sunday night it was reported to be contained; but fire crews lost control, and by midday Monday it had taken off, pushed by high winds (which *are* typical of the season) across an area called Doney Park. I think of Doney Park as the place where people move to live an Old West rancher lifestyle. Single-family homes sit on large plots of land, and many people have horses, chickens, or other livestock.

The fire came up so fast, with the status changing from "set" to "go" rapidly, that people had little time to prepare to evacuate. Though no deaths were reported, Flagstaff social media was filled with stories of people who barely made it out in time, including one young family with a newborn baby who, waking up from a collective afternoon nap, had to race for their vehicles before their house was engulfed in flames. Shelters were set up in the city for the human residents of the area, but they couldn't accommodate nonhuman evacuees. A large public call went out—here, again, on social media—to help people find temporary homes for their horses, pigs, and other domestic animals, who had been rushed out of their pens at the onset of the flames.

The Tunnel Fire eventually burned itself into a national monument, Sunset Crater, where fire crews were able to contain it without any further fear of structural loss or major habitat damage. But it was followed less than two months later by another fire in the Kachina Peaks Wilderness, the Pipeline Fire, which burned another 20,000 acres. The Pipeline Fire consumed fewer human-built structures, but it scorched well-loved hiking trails, key areas of Flagstaff's watershed, and culturally important sites for the Diné, Hopi, and other Southwest Indigenous tribes.[25]

These events—both shocking and saddening—left Northern Arizona residents acutely aware of how quickly fires can turn well-loved sites to cinders. My own family now lives in a house on the edge of forestland, about twelve miles outside of Flagstaff. In the dry season, neighbors keep a close lookout for smoke plumes on the horizon, and when one appears we gather to watch and speculate, until we have news that fire crews have it under control. This is the first place I have lived where, during fire season, it feels prudent to keep a go bag, prepacked with our irreplaceables, in case of a sudden evacuation order.

And then, there are floods. Floods aren't new in this part of the world; flash flooding during the monsoon season is part of the ecology of the desert Southwest. Intense summer rainfall exceeds the soil's absorption capacity and runs across well-worn paths, turning dry channels into rushing rivers.

But in recent years, with more of the year's precipitation shifting to the summer monsoons,[26] floods have combined with high levels of urbanization and short-sighted city planning to create major problems for the city of Flagstaff. In the summer of 2021, following the neighborhood-skirting Museum Fire of two years prior, what was described as a "200–500 year rainfall event" sent fire debris down the slopes of the mountain through East Flagstaff, turning the streets into muddy, rushing waterways, carrying burnt logs and branches and other debris.[27] Evacuations weren't required, but friends reported water up to a foot deep around their houses. One image of the flooding, which went viral on social media, showed the floodwaters lifting a parked Toyota Prius and pushing it down the street like an ungainly boat, until it crashed into another parked car.

In human terms, the 2021 flood alone damaged some eighty-eight private properties and thirty-three public buildings, with a cost estimated at around $2.3 million.[28] In ecological terms, too, these events have ongoing potential for damage. Though flooding itself might be "normal," the kind of flooding that occurs after severe wildfires can degrade water quality,[29] and interactions between fire and other landscape dynamics can lead to unexpected and nonlinear outcomes that cross landscapes and scales.[30] These repeated events speak of a more-than-human landscape in a state of precarity—and how this meshes with the precarity of other mostly human systems is where we turn next.

Making Meaning Amid Multiple Emergencies

Plague, fire, flood. But there's more.

One might think that, amid all of these disasters, Flagstaff would gain a reputation as a dangerous place to live and see a mass population exodus. Not so. At the peak of the pandemic, the "Zoom boom" drove a huge increase in demand for houses in Flagstaff. Tech workers ("from California," the story goes) were seeking beautiful and lower-density areas from which to work remotely. With Silicon Valley salaries—and often a great deal of cash, accumulated from equity in California's overpriced housing market—they swooped in and purchased homes sight unseen.

Flagstaff housing was already expensive and in short supply. The ability to open new areas for construction is constrained by the fact that the city is surrounded by public lands; and much of the available housing is second homes or Airbnbs. The trails, the forests, the local ski resort, and the proximity of the Grand Canyon make the area a desirable vacation destination. The demand for housing had already been on the upswing before the pandemic, with the expansion of Northern Arizona University's student population driving an increase in the need for (and price of) rental units. The combination of the Zoom boom and low interest rates meant that prices of real estate increased 21 percent in 2021 alone.[31] Houses that went on the market were scooped up within days of being listed, often with bidding wars driving the price above asking. The price of rental units increased apace.

The city of Flagstaff has declared a housing emergency, citing the fact that, over a period of ten years, the median sale price of a home rose 53 percent, while area median income increased by only 14 percent.[32] In local public information groups on social media, families regularly post requests for help with housing. Commonly, they cite evictions due to economic hardship and report difficulty finding well-maintained places of a size suitable for a family on a local salary. Many have taken to living in the surrounding national forests, setting up tents or vans on public land where camping is not restricted. Imagine: Amidst unseasonable weather and a global pandemic, what would it be like to lose your shelter?

The past three years in Northern Arizona have seen a pileup of unprecedented or once-in-a-lifetime events. Taken together, there is a fictive quality to this, reminiscent of Naomi Oreskes and Erik Conway's *The Collapse of Western Civilization*, in which they imagine how a series of theoretically low-likelihood occurrences might build upon one another to lead, effectively, to a climate apocalypse.[33] Living in Northern Arizona in this period—or perhaps, living anywhere in the contemporary moment—feels a bit like experiencing this sort of chain of events: a low-grade, distributed apocalypse. "Extreme" or unseasonal weather comes regularly, punctuated by other unprecedented events. It happens not with the definitive narrative progression of big-screen climate disaster films, but dispersed just enough across groups of people, and across space and time, that it's easy to talk oneself into thinking that things will get "back to normal," projecting a storyline of certainty into an uncertain future.

But what would "normal" look like—and who decides? The interpretation of extreme events is closely related to how people imagine the past, and the future, of life in Northern Arizona—and nowhere is this clearer than in discussions of housing. While some argue that what's needed, desperately, is high-density, climate-resilient urban housing, others say that Flagstaff's population has already grown too much; they gesture at the idea that "some people" should make the rational choice to leave, if they can't afford to stay. Still others retort that this strategy would decimate Flagstaff's laboring class: the mechanics, the plumbers, the teachers, the service industry. They ask, Who would clean the Airbnbs, or serve the tourists, if the town were entirely populated by short-term rentals?

And not far behind these conversations lies the simmering reality that for Diné, Hopi, and many other Indigenous groups, Northern Arizona is "home" not just by happenstance. Rather, staying in Flagstaff is a middle ground between the affordances of urban life (schools, piped water, grocery stores, jobs) and a tenacious connection to history and sovereignty. Extreme events represent a threat to property and livelihood, but for Indigenous people in Northern Arizona, as in other places, changing climates are simply another in a long history of existential threats.[34]

In the unresolved public debates around extreme weather, fundamental differences about what it means to live in community—or, what exactly *are* the relations uniting people who live in the same space—are at the forefront. So, too, are relations to the more-than-human world, which are at play in conversations about neighborhood zoning, floodwater management, hiking trails, and respect for culturally important Indigenous sites. Thinking with a material phenomenology that combines the situated and intersecting experiences of materiality and meaning making offers a starting point for understanding why climate and other environmental problems seem so intractable, and for tracking how, in times of uncertainty, particular configurations of collective active and decision making do, or do not, emerge.

Climate Is Planetary, but Also Particular

In each of the places I have described—Bolivia, Mexico, Arizona—the unseasonal weather patterns linked to climate change are experienced differently. These differences result not just from shifts in precipitation, temperature, or extreme events, but also from how climate layers into other forms of experience, like politics, livelihoods, faith, identity, and infrastructure. Understanding this layering requires a toes-in-the-dirt approach, starting from localized experiences of both the human and more-than-human. The view from a locality—or the patch—is about understanding the simultaneity of these experiences. Situated knowledge offers a point of entry for understanding how people experience the kinds of rupture represented by unseasonal weather. It also helps in making sense of what people hold onto as they, necessarily, forge ahead through uncertain times.

For those of us (anthropologists and others) who are in fields concerned with meaning making, the globality of climate politics, funding streams, and media is important—and also often confounding. Under some circumstances, an emphasis on globality can get in the way of seeing "the local" by encouraging researchers to minimize, or alternatively overinterpret, the specificities of what happens on the ground. This is the kind of pitfall that Mike Hulme has described as "climate reductionism."[35] Uncritical acceptance of these reductionisms can lead to misreadings of the processes that are actually occurring in place.[36]

Drawing from Anna Tsing's writing on globalization, we instead might think of knowledge about climate change as a "global universal," an idea with broad geographic reach and worrisome hegemonic potential.[37] But as critical climate scholars show, the uncritical application of universality is often an indicator of underlying heterogeneity: Forces of disjuncture roil within conjuncture and calls for unity mask longstanding power struggles or minimize historical injustices. Claims to universality also draw the focus away from all but a narrow range of nonhuman actors—among them, the kind we describe here as feral.

A patchy analysis of climate change puts focus back on these kinds differences—not as ancillary to the planetary phenomenon, but rather as constitutive of it. Globality is only one way of understanding climate change; another is the view from a patch.

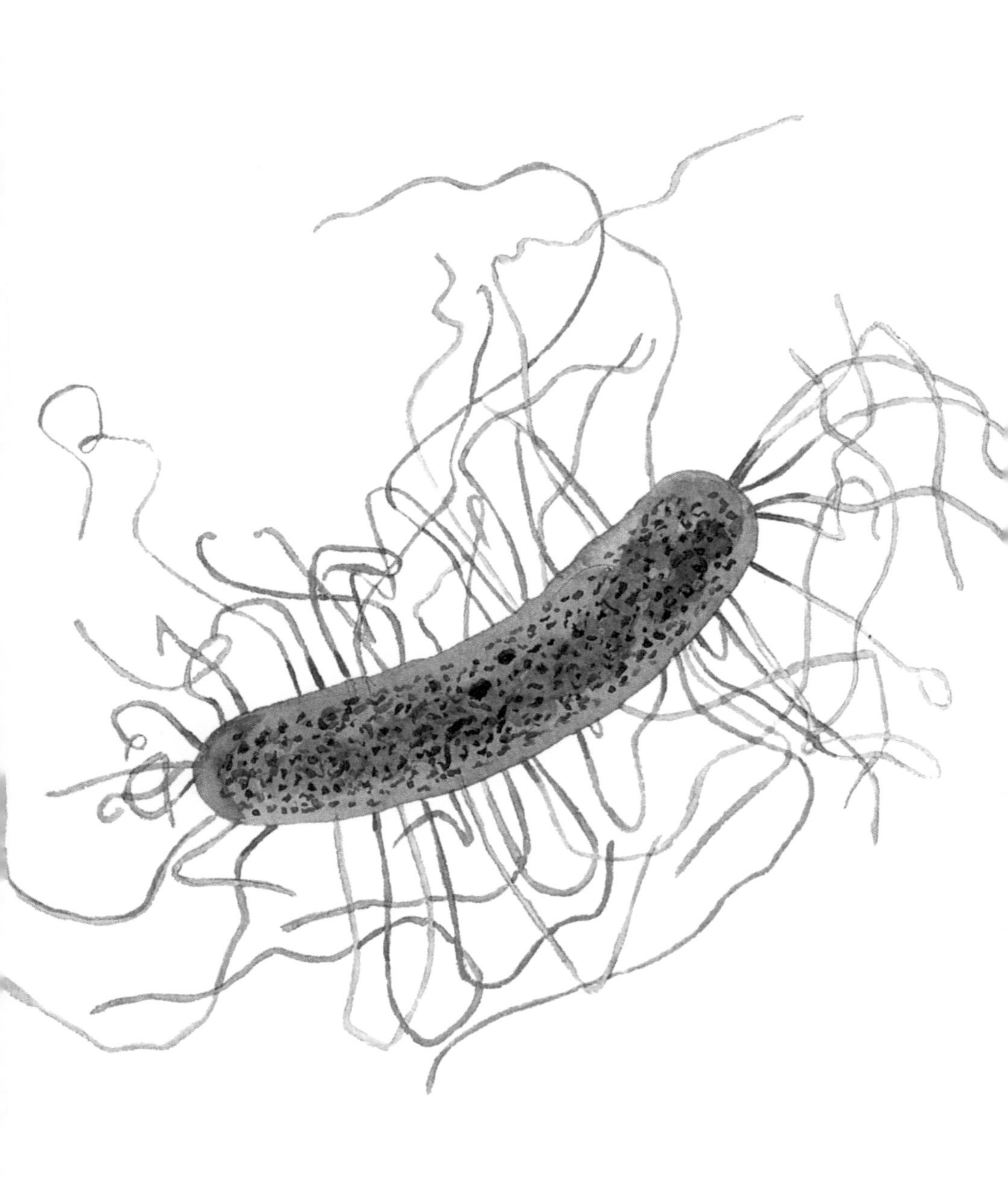

SIX

How to Detonate an Anthropocene

How do anthropogenic patches gain a planetwide force? There are many good answers to this question. Following the discussion of climate change in the previous chapter, we start here with the circulation of carbon in air and water, perhaps the most widely recognized patch-to-planet example. Anthropogenic carbon dioxide, produced by burning fossil fuels in patches of industrial infrastructure, enters the atmosphere, eventually circling the planet. Plants use carbon dioxide in their metabolism, reducing anthropogenic carbon dioxide; oceans are also an important sink. But they cannot eliminate the overall increase in atmospheric carbon dioxide produced by the burning of fossil fuel; thus the planet warms.

Though geologists have steered away from using concentrations of atmospheric carbon as the defining geologic signature of the Anthropocene,[1] the change in carbon dioxide concentrations remains an important referent for Anthropocene processes. This change, indeed, raises lots of further questions that can help us delve deeper into Anthropocene-making processes. Under what conditions did those fossil fuel–burning infrastructures get there in the first place—and why have they been allowed to proliferate even after most everyone realized that they threaten more-than-human life? Meanwhile, what about all the other environmental catastrophes surrounding us, from the extinction crisis and the pandemic to toxic poisoning? To address these questions, we need more ways to understand shifts in patch-to-planet relations. This chapter introduces Anthropocene "Detonators" to offer a history of the Anthropocene. These Detonators draw political economies into understandings of the Anthropocene, while also showing how an array of nonhumans get involved at a planetary scale.

How Did Those Systems-Changing Infrastructures Get There?

If, as this book argues, imperial and industrial infrastructures have been particularly implicated in producing Anthropocene hotspots—that is, sites of systems rupture—then it seems appropriate to ask just how all those infrastructures came into being. What kinds of building programs were responsible for them? The work of environmental historians and social scientists has been particularly useful in answering these questions, and this book follows them to identify four kinds of infrastructure-building programs that have gained planet-changing importance through their systems-changing patches.

First, beginning at the end of the fifteenth century, programs of *multispecies invasion* have marked European expansion into new territories. Multispecies invasion killed and displaced native communities and ecologies, allowing the formation of the programs that today we call settler colonialism. The late fifteenth-century invasion of the Americas set a model that has been continued, across centuries, offering a too-familiar reenactment in places such as Australia and New Zealand. In each of these places, earlier human residents were brutally murdered, sickened, dispossessed, and disinherited, as the invaders set up their own projects for settlement. Whole suites of flora and fauna were introduced, along with their microbiomes, and these introduced species carried out much of the work of invasion for the settlers.[2] Massacres, diseases, and forced migrations opened landscapes for settler remaking and reuse.[3] The continuing introduction of species and remaking of landscapes to address elite fantasies and offer material support to imperial missions, such as "rural development," is part of this legacy.

Second, from the very beginning of the seventeenth century, programs of European imperial expansion have introduced infrastructures tied to missions of governance and trade for the metropole. We might call these *terraforming governance* programs, following the term used in science fiction to describe the settlement of new planets.[4] Imperial armies were distributed, militarizing trade; terraforming followed, remaking the earth's surface. In places where water and land overlapped, such as the coasts of the Indian Ocean, harbors were dredged, canals drained swamps, and roads and eventually railways were built to extract goods to send to imperial centers. Plantations were put in place for the cheap and efficient production of plant material for what became European industry.

Third, the political economy we call *industrial capitalism* took off from such arrangements in the late eighteenth century, as the windfall profits of New World plantations supported the invention of industrial technologies. At this same time, "free" labor became available to work in the metropole's factories through the displacement of people from the countryside. Industrial capitalism crawled out of the belly of "war capitalism," the term historian Sven Beckert uses to describe New World plantations using enslaved labor.[5] Since this time, this political economy has become more powerful, shaping state rule, property law, nature, and war to serve the accumulation of wealth by elites. Capitalism "makes nature cheap," in the words of sociologist Jason Moore.[6] Moore argues that the success of capitalism has been made possible by allowing corporations to externalize environmental costs. This means that the companies that use natural resources do not need to consider the environmental destruction left in their wake. Industrial capitalism has become a vast infrastructure-building program. Capitalism makes investments equivalently liquid, whatever their social and ecological effects. Thus, investors are encouraged to terraform distant lands for their projects, entirely disregarding the effects of these projects on local people and ecologies.

Fourth, these three programs of infrastructure building were combined and accelerated in the twentieth century, especially in the period since World War II that Anthropocene thinkers have begun to call the "Great Acceleration." The acceleration marked by most of those who use this term refers to a quantitative uptick in earth systems disturbances as well as human populations and activities.[7] These numbers come into being through a political economy and ecology, and these are key to infrastructure building in this period. This is the period of the Cold War and rising U.S. hegemony in world politics. It is also the period of decolonization, in which nation-states began to cover the earth. As national elites took control, under the shadow of the great powers, they learned to sponsor all the infrastructure-building programs inherited from their colonial pasts. Thus, multispecies invasion, terraforming governance, and industrial capitalism became democratized in the sense that national governments all over the world could use them to expand their territories, govern or squash unwilling people, and destroy their ecologies for capitalist investors. A great orgy of destruction has followed across the world, and we live in its wastes and ruins.

Each of these programs of infrastructure building could be seen as an origin story for the Anthropocene; each created patches in which state changes have created challenges for more-than-human life. Their programs changed—and continue to change—the kinds of infrastructures being built in particular times and places. They do not conjure uniform results across the globe. But each lends a configuration of force—some blend of military, economic, cultural, or legal-political justifications and incentives—to a set of projects. Even scattered across time and place, the cumulative force of the infrastructure-building program authorizes its proliferation—and its detonation of Anthropocene ecologies.

Here's a small but vivid example of that cumulative force. In *Running Out*, anthropologist Lucas Bessire asks why settlers used pumps and deep wells to drain the aquifers in Kansas, even after they knew that, by doing so, they would run out of water.[8] To answer this, he piles up local and regional histories. From a zoomed-out perspective, one can identify that kind of drilling as part of an infrastructure-building project. It occurred to the early settlers because they knew of such projects in other places. It was part of the taming of the U.S. landscape, a terraforming governance project, in the language above. It became common sense to the settlers, and then it seemed necessary to keep up with one another as well as with farmers in other settler areas. The common sense of their success—together with its backing by government authorities, banks, local councils, agricultural equipment companies, commodity buyers, property and water laws, God's will, and more—reinforced their efforts to keep such infrastructures in place, and rebuild them, long after the news was out about the oncoming desertification of the area. The conjuncture is the force of infrastructure-building programs: the force that allows pumps to keep draining aquifers even against all good sense.

How do infrastructure programs detonate the Anthropocene? Detonation describes a particular mode of temporality in which historical events, while occurring in a chronology, create system ruptures, which operate beyond chronologies because they form new modes of relationality. This is a mode of temporality that—unlike the golden spike—does not demand a single answer to the question, "When did the Anthropocene begin?" Systems shifts can be multiple and overlapping. A single systems shift may have many iterative points of initiation and deepening. We suggest start dates for the infrastructure-building programs we describe, but not end dates. Each continues now into an indefinite future.

Anthropocene Detonators Make New Worlds

To detonate is to cause to explode. Each of the infrastructure-building programs introduced above has been responsible for an explosion of changes across the critical zones—that is, the zones of life—of the earth's surface.[9] These changes happen in patches, but they are not isolated. Returning to the example of aquifer tapping above, deepwater drills have been established in many places across settler landscapes, and they connect to one another through related forms of material, ideological, and institutional backing. Yet the space-time coordinates of these infrastructure-building programs are difficult to map. Consider a wildfire, spreading unevenly across the landscape, following winds up ravines or smoldering through underground charcoal to pop up, unexpectedly, in an entirely different place. It is impossible to map the full extent of a wildfire while it is still burning. And even then the shape may be irregular in both space and time. Sometimes the fire races along narrow corridors between unburned swaths of forest; even where it runs, some stands are spared while others erupt in flames. Sometimes the fire leaps back into action after it has seemingly extinguished itself. These very problems of containment make wildfires a force to be reckoned with. The metaphor suggests how imperial and industrial infrastructures and their feral effects proliferate. Like a fire, they race along corridors, sparing occasional stands while obliterating others, and leaping to new places just when the whole program seems dead. Just because they cannot be contained by a conventional timeline or map does not mean these projects are insubstantial: They are material, insistent, and hard to keep down, rather like a wildfire.

One way to convey the force and extent of these infrastructure-building programs is to introduce the series of hand-drawn landscapes Feifei Zhou designed for *Feral Atlas*. We labeled these with the following shorthand terms, corresponding to the four infrastructure-building programs outlined in the previous section: *Invasion, Empire, Capital,* and *Acceleration*. These landscapes are composed through the collaging of historical referents. They skip across time, space, and conventions of representation, juxtaposing diverse referents. Through this juxtaposition, we aimed to show the wildfire-like force of these programs, as they irregularly incinerate varied times, places, and modes of knowing the past.

Each of these landscapes represents a world-making perspective associated with a particular set of infrastructure-building projects. The program of *multispecies invasion* is represented as a long horizontal journey: a spreading smear of continent-crossing invasion practices. Feifei Zhou worked here with contributions from two Indigenous artists to tour the new world created by invasion (see discussion in Chapter 11). Viewers are taken to scenes that range in time from the late fifteenth-century invasion of the Americas to today's North American settler colonialism, and range in place from early twentieth-century Australian battles between Aboriginal families and police over the mining of ancestral lands to the contemporary cattle invasion of the Amazon.

The program for *terraforming governance* is shown as the view looking down from a hill, a colonial surveyor's perspective. The view reveals the infrastructure-building programs of empire, from plantations to water management. For this landscape, Zhou worked with Ghanaian British artist and architect Larry Botchway, who has drawn a landscape of oil palm cultivation by smallholders, who have been forced to adopt the practices of large-scale plantation agriculture because of contracts to the buying company (see Figure 51 in Chapter 11).

Industrial capitalism as an infrastructure-building program is shown as a grid, representing the incessant commodification of everything. In Figure 25, you see the grid form of industrial production and the grid form of supermarket consumption side by side, mimicking each other. But here there is decay: Smog circulates above the city; non-metaphorical wildfires burn; and acid mine drainage, sewage, and trash seep into the water. The city and the countryside join in a tight relation of capitalist investment, forming what historian William Cronon calls "second nature," the nature made by capitalism.[10] But there is also a "third nature" here, the feral effects of capitalist infrastructure.[11] Pests and pathogens flourish; weeds spread. The power plant burns; the cargo ship keels over; city traffic is in a snarl. A demonstration breaks out at the heart of the city, and rebels haunt the coast. Such effects are not just associated with the breaking down and wearing out of infrastructure; they are effects of the infrastructure itself.

The Great Acceleration, as a Detonator landscape, takes us underwater, looking up at a clot of marine garbage. The waste and poisons of the Anthropocene surround the viewer; there is no place to run. The world has become claustrophobic. Here Zhou's collaborators are Filipinx art-

FIGURE 25. **Detail of *Capital.***
FEIFEI ZHOU

FIGURE 26. ***Invasion.***
FEIFEI ZHOU WITH NANCY MCDINNY AND ANDY EVERSON

ists Amy Lien and Enzo Camacho, who offer a figure of disorder and perversity (see Figures 48 and 49 in Chapter 11). This is the catastrophe of the Anthropocene, which suffocates us from every side.

By associating field reports with each of these landscapes, *Feral Atlas* shows patch-to-planet connections that develop through these great infrastructure-making programs. Meanwhile, the diversity in the referents within each landscape painting offers insights about the specificity of the infrastructure-building program. Some examples from these landscapes should clarify how these Detonators generate planetary effects. For the discussion here, we choose Invasion to get readers started on thinking across space and time with Detonators, and Empire to explain the force of colonial interventions. We hope readers will continue the journey by looking at the full set of Detonator landscapes in *Feral Atlas*.

Multispecies Invasion Characterizes Settler Colonialism Across Continents

European expansion since the late fifteenth century has been characterized by multispecies invasion, that is, the use of other-than-human species as part of the project of invasion (Figure 26). Europeans traveled with a portfolio of plants, animals, and disease organisms, and human

invasions could not have succeeded without this multispecies gang. Disease organisms from Europe turned the American continents into a terrifying zone of illness, resulting in a 90 percent population decrease.[12] Native foraging ecologies were destroyed as animals and plants long associated with European-style disturbance rode roughshod over the American continental environment. Ever since, Native Americans have had to navigate what Potawatomi scholar Kyle Whyte calls "our ancestors' dystopia."[13]

Scholars have argued about whether the Europeans used multispecies invasion in the New World purposely or without design. For our purposes, the question of intent is less important than the consequences. It is clear that some Europeans handed out blankets *in order to* transmit smallpox, for example, and also that enslavement, land grabs, and malnutrition added to the deadliness of European diseases. But it also seems likely that Europeans took advantage of multispecies invasion where they had not thought to plan it. Virginia Anderson has described how European settlers on the eastern seaboard of North America allowed their cattle and pigs to roam freely, destroying Native people's gardens.[14] When Native people dared to hunt them, the settlers razed whole villages in return. Settlers benefited from the mix of designed disturbances of Native life, of unexpected disturbances to use to their advantage, and of merely not caring when the organisms they brought with them destroyed Native people and ecologies.

Thinking through multispecies invasion as a program allows us to consider the global and planetary impact of what otherwise might seem isolated instances. Consider three separate examples, each from a separate continent and a separate period, yet connected through the infrastructural program of invasion. Rosa Ficek describes the introduction of exotic pasture grasses in early twentieth-century Panama by colonial settlers who found such grasses ideal for cattle raising.[15] By keeping tree seedlings from becoming established, the grasses themselves have held back the forest ecologies on which Native people depend. Native communities have been forced to retreat. Colonists knew that the grass was part of their expansion: Ficek cites a Panamanian geographer who speaks of the "conquering grass." Panamanian writers refer not only to its pleasures (feeding cattle) but also to its terrors—which include pushing out even settler farmers since the grass overruns their farms. It has a force beyond their control. The grass allowed ranchers to become wealthy property owners. It spread their territories for them. Grass has provided a kind of nonhuman labor: a force for the transformation of landscape. It participated in the invasion of indigenous ecologies and the displacement of Native communities by creating a world for the expansion of cattle ranching.

Meanwhile, in Australia, settlers were also busy introducing new species. In 1935, they brought the Central and South American cane toad as a potential predator for the beetles that infested settler-planted sugar cane.[16] The cane toads never got close to the beetles; instead, the toads spread across northern Australia, eating a huge variety of insects along with frogs, small mammals, human waste, and petfood. Of course, other animals also tried to eat them. But the poison glands in the necks of the cane toads are powerful and effective: Many animals that have tried to eat cane toads, such as snakes and lizards, have died before they could even swallow a bite.[17] They died before they could learn to avoid the toads. Since the toads are legion, so too have been the deaths. This has been an important problem not just for multispecies ecologies but also for Aboriginal communities for whom many of the dying animals represent not only important foods but also precious kin.

Yolŋu Aboriginal artist Russell Ngadiyali Ashley created a painting

for *Feral Atlas* to show the effects of cane toads on his extended family (Figure 27).[18] The painting centers on a species of goanna lizard, which was both a livelihood staple and an ancestral animal for his clan. The goannas are being protected by warriors with spears, surrounded in ceremony by men and women dancing and crying for their kin. Outside the ceremonial circle, we see a number of large toads clearly on the move. A particular species of goanna has disappeared from his area, Ashley explains; they have been poisoned by the cane toads that first appeared in large numbers in East Arnhem Land in the early years of this century. Yolŋu Aboriginal people celebrate their goanna kin, and mourn for them, but have not been able to stop the toads, which carry on the violence and displacements of settler invasion all by themselves. "Before, goannas were here forever," Ashley laments in his commentary. This formulation of ecological time and loss offers another patch-based perspective on systems rupture (see Chapter 4) as Ashley explicitly figures the Anthropocene as a violent clash of more-than-human temporalities. In fact, the painting precisely depicts this devastating temporal encounter. On the left side, we see the ceremonial realm in which goannas exist "forever"; on the right, we see the cause of ceremonial tears and spears: the new toxic "stranger" moving across Yolŋu lands, decimating ancestral histories and intergenerational continuities in its wake.

Historian Elizabeth Fenn starts much earlier in Anthropocene invasion history in her documentation of a Mandan village along the Missouri River, now part of North Dakota in the United States.[19] This village experienced a series of population collapses due to the diseases brought by Europeans in their invasion of the Americas. The first collapse she finds was in the late 1500s. But the village worked its way back—until the mid-1700s, when the population collapsed again. Smallpox—spread by trade, war, and people newly mounted on horses (also a European import)—was probably the culprit for this round of disease. The collapse of the village's population made the survivors vulnerable to attacks from their enemies. Eventually, the remaining survivors agreed to move so they could join with allies.

Traces of the abandoned village still survive today, as seen in Figure 28. However, the region the villagers abandoned has been put to new settler purposes. Oil shale extraction occupies the region. A dam has

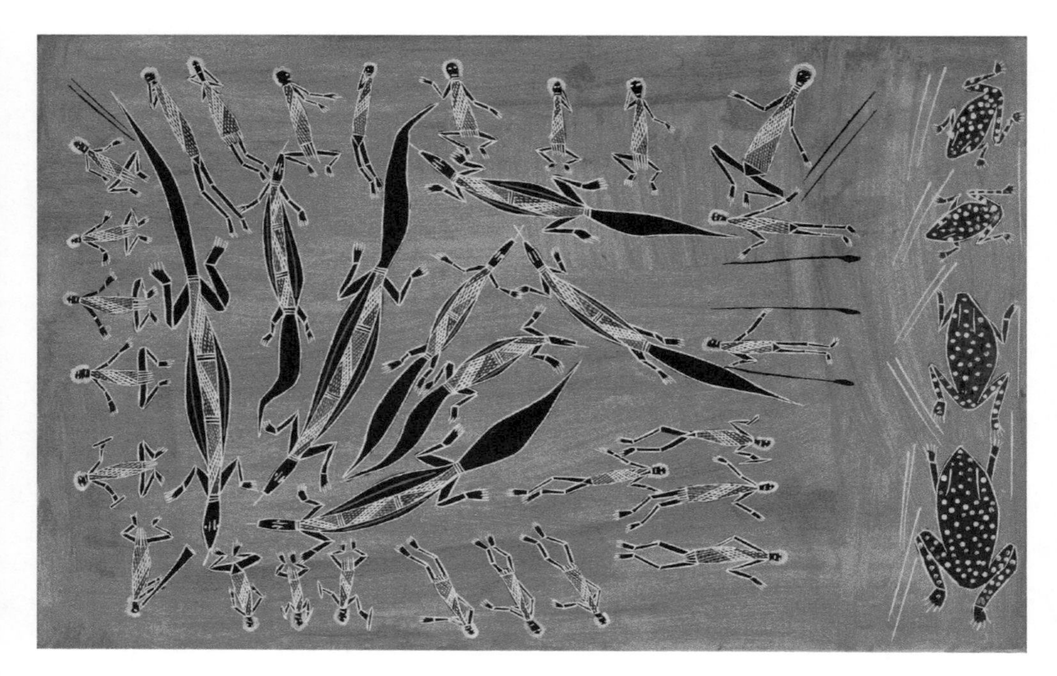

FIGURE 27. ***Yätj Garkman*** **(Evil Frog).**
RUSSELL NGADIYALI ASHLEY

flooded the lands once used for Mandan villagers' farms. The dam endangers native fish, such as pallid sturgeon, which were important to Mandan communities. Settler diseases killed and displaced Mandan villagers; now settler projects replace their landscape.

Each of these reports (grasses in Panama, cane toads in Australia, and diseases in North Dakota) stands alone as a testimony to the force of multispecies invasion. However, considered together, the importance of this program for both world history and planetary biodiversity can be better appreciated. Unfortunately, multispecies invasions have become accepted as an ordinary, normalized part of the human condition. The human and nonhuman parts of the invasion have been analytically separated, as if they had nothing to do with each other, perhaps because settler domination seems so banal that it is rarely remarked upon. Biologists, buried in the lives of plants and animals, too often discuss them in isolation from human projects of introduction. Scholars in the environmental humanities are inclined to respond with bleeding-heart sympathy for invasive species and criticize the discussion as xenophobic. In the

FIGURE 28. **Aerial view of Yellow Earth (or Double Ditch) Village.**
MICHAEL FROHLICH

process, these scholars further segregate human and nonhuman invasion projects, accepting their isolation from each other, as if the species transfers involved in human invasion were not a world-historical force. By identifying multispecies invasion as an infrastructure-building program that detonates the Anthropocene, we can restore the connection. "Introduced" plants and animals are introduced by humans—and not, for the most part, by little old ladies slipping seeds into their handbags before flying home from abroad, as popular imagery has it. The most troublesome introductions have formed part of human projects of conquest and commerce. In every case, the connection bears watching.

As we consider the more-than-human making of settler colonialism, we gain a new appreciation of Indigenous peoples' kinship relations with nonhumans. As Dakota scholar Kim TallBear explains,

> The decimation of humans and nonhumans in these continents has gone hand in hand. When one speaks of genocide in the Americas it cannot be understood in relation to the European holocaust, for example, that is seen as having a beginning and an end, and which is focused on humans alone. Our genocide in the Americas included and continues to include our other-than-human relatives.[20]

The deaths are connected. "Decline of the sturgeon has corresponded with decline in sturgeon clan families," explains Anishinaabe elder Kenny Pheasant, quoted by Kyle Whyte.[21]

Multispecies invasion has been associated with extinctions as well as the decimation of both humans and nonhumans. But what do these losses have to do with the geology of the planet? Each of the three field reports we have introduced has important geological impacts; the work of geographers Simon Lewis and Mark Maslin is helpful in understanding how.[22] Lewis and Maslin noticed that the "Orbis spike," a 1610 dip of carbon dioxide levels in the earth's atmosphere as recorded in Antarctic ice, occurred as Europeans had consolidated their hold on the Americas—causing millions of human deaths in the process, particularly through disease. The deaths of Native Americans caused the regrowth of forests where farms had once flourished. Lewis and Maslin argue that this forest regrowth could have caused the drop in atmospheric carbon dioxide. They also note that the coldest period in Europe's Little Ice Age occurred soon afterwards, perhaps corresponding to the force of multispecies European invasion in the Americas. The introduction of European plagues had an impact so profound that it can be seen today in layers of rock. This explanation creates a truly planetary climatological story from regional historical records.

The more recent cases of multispecies invasion discussed here also have the potential for planetary effects with geological signatures. The introduction of cattle and pasture grass in Latin America has had a strong impact on the size and health of forests, which in turn has affected the carbon dioxide composition of the planet's atmosphere. Pasture grasses cannot metabolize carbon dioxide to the extent of the forests they replace. The loss of Latin American forests continues to be driven primarily by conversion to pastureland.[23] Through both the release of carbon—and loss of forest systems that absorb carbon—this process contributes to climate change, effectively the inverse of the 1610 carbon minima that Lewis and Maslin identify.

What about goannas? J. Sean Doody and his colleagues have shown that these predatory lizards play a serious role in ecosystems engineering in northern Australia. Goannas make deep communal burrows, which form refuges for many other species. Doody's team looked in goanna burrows and found

> snakes, geckos, skinks, monitors, frogs, toads, scorpions, centipedes, beetles, ants, and a marsupial, sometimes one at a time, other times in great numbers. One warren even contained 418 individual frogs. Overall we found 747 individuals of 28 species of vertebrates in just 16 warrens.[24]

These findings speak of a complex web of subterranean multispecies interactions—which in turn raise questions about how the decimation of goanna populations will influence surrounding ecologies. Research at the interface of landscape ecology and wildlife biology has only recently begun to document the extent to which animal interactions influence larger processes of nutrient cycling, or what ecologist Oswald Schmitz and colleagues refer to as "zoogeochemistry."[25] The zoogeochemical consequences of the loss of goannas from Australia's ecological web, then, are still unfolding.

The feral dynamics described in each of these three field reports have larger geological consequences. Taken together, even these three start to offer a prescription for planetary trouble. But now imagine these, and more: The consequences of multispecies invasion have been enormous. The patches in which they develop have considerable autonomy, but also a strong thread of connection. Piled upon each other, they detonate the Anthropocene.[26]

Each of the Detonators we propose is similarly powerful. In explaining imperial governance, we turn back to the landscape Feifei Zhou developed for *Feral Atlas*.

Terraforming Governance Created Experimental Ecological Machines

As Europeans asserted control over larger and larger parts of the earth's surface, they took advantage of local knowledge systems, hierarchies, and agroecological formations to build imperial projects. In many cases they distorted local systems even as they regularized and spread them. In the process, they created "experimental" systems, in Jerry Zee's sense of the term, that might make imperial governance possible.[27] These systems were machine-like to the extent that they were standardized to create regular results, and to the extent that they were transported in these standard forms to other places around the world to replicate those

FIGURE 29. ***Empire.***

FEIFEI ZHOU WITH LARRY BOTCHWAY

results. Art historian Jill Casid uses the term "machine" to describe Caribbean plantations, which, she argues, developed regular material and semiotic patterns in the production of colonial commodities.[28] We follow this use of machine to describe a larger mix of earth-modifying projects designed to facilitate colonial governance and its associated extraction economies. The global and planetary effects of imperial terraforming are directly related to these machine-like qualities, which allow each project to serve as a model for similar projects across the world.

Consider some examples from Zhou's *Empire* Detonator landscape in *Feral Atlas* (Figures 30–33). Figure 30 shows one corner of this landscape; it represents the taungnya system for growing teak developed in British colonial Burma in the nineteenth century. (George Orwell is shooting an elephant in the upper right corner.) Taungnya was developed as a colonial mimicry of local agroforestry practices in Burma's hills, in which multiple crops, from annuals to trees, were planted together, allowing layering and harvests over time. In the British colonial version, local communities were allowed to plant subsistence crops (in the image, cassava) in exchange for taking care of the young trees. This provided labor for the colonial commodity. In contrast to local forms, however, the goal was to produce a monocrop teak plantation by the time the trees matured. Subsistence crops were allowed only when the trees were young—and needing human care. In the early twentieth century, the system broke down, in part because monocrop plantations nurture pests, here the teak beehole borer. But meanwhile taungnya was exported to other places, including India, Kenya, Nigeria, Indonesia, and Thailand.[29] British experiments in Burma created a terraforming system that could be exported with standard features: an agroecological machine.

Another agroecological machine with an important historical footprint was the forced Cultivation System of the Netherlands East Indies, as deployed in nineteenth-century colonial Java (Figure 31). (Anthropologist Clifford Geertz describes this system in his book *Agricultural Involution*.[30]) Javanese peasants worked irrigated fields, growing rice for their subsistence. Dutch colonial authorities, working through local elites, took advantage of the irrigation system to demand that these peasants alternate rice with their favored colonial crop, sugarcane. Since they were not compensated for growing sugarcane, this system put consider-

FIGURE 30. **Detail of *Empire* (1).**

able pressure on already struggling peasant families, who responded by having more children to increase their labor power, which put further pressure on the system.[31] The result benefited the metropole, however, bringing the Netherlands from the brink of bankruptcy into a growth economy. Here was another way to bring labor into the production of colonial commodities without cost to the metropole.

The most infamous system for bringing labor to colonial crop production was the kidnapping and enslavement of people from Africa to serve as laborers on New World plantations. Again, local systems of hierarchy and exchange were tapped—and distorted—as the Atlantic slave trade gathered laborers through warfare. If you turn back to Figure 29, the full version of Zhou's *Empire* landscape, you can see Ghana's coastal slave fort, a military delivery structure for enslaved people, next to the ocean on the far right. Such fortifications suggest the violent distortion

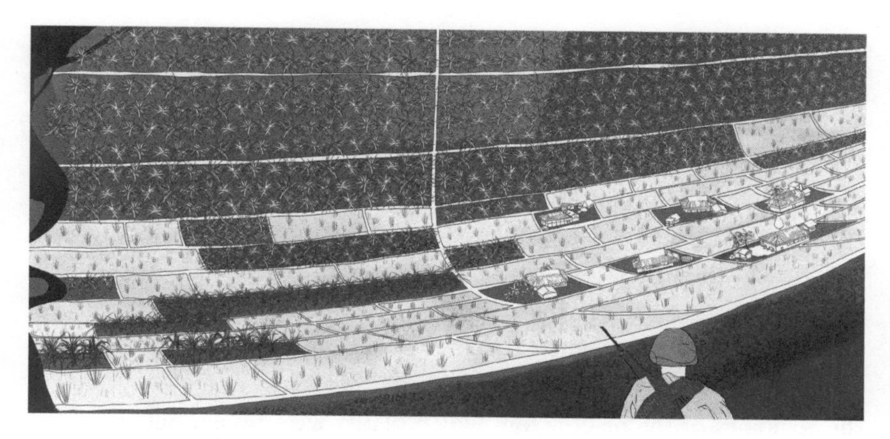

FIGURE 31. **Detail of *Empire* (2).**

of local African hierarchies that occurred as Europeans incited the kidnapping and export of more and more people. In this image, labor is being delivered for sugarcane production in the New World. Other plantation crops were grown in the Americas through this coerced labor system, including rice, which depended on the knowledge of enslaved West African people, who had grown rice in West Africa.[32] Their knowledge was funneled into an experimental system, a model for imperial terraforming.

Plantations were not the only imperial terraforming system. The *Empire* landscape also shows another important ecological experiment: the separation of land and water to make each more governable as well as more efficient for colonial uses. (Note that this landscape painting includes both colony and metropole, so viewers get a glimpse of water management in each.) Harbors were dredged for the convenience of imperial shipping. Swamps were drained with canals, which could also be used to transport goods. Dams collected water for the irrigation of commercial crops (Figure 32). In building these water management infrastructures, colonial engineers borrowed heavily from indigenous state projects in the Global South, especially in South and Southeast Asia, where water management had been a state project for centuries. As these projects were copied, however, they were also remade for long-distance rule as well as for transport to new locations. Often, colonial innovations reverberated in the metropole. Historian David Biggs documents the use of construction with interchangeable parts in the Mekong Delta of French Indochina to build bridges and railway lines; these inspired building design in France, including the Eiffel Tower.[33] (The bridge in Figure 32 is an example of this kind of construction.) The interchangeable parts of French colonial construction are a vivid figuration for the module-like program of imperial terraforming.

One other use of water management is worth mentioning: the botanical garden (Figure 33). Plants from all over the colonies were brought together in the artificial environments of the botanical garden, where they served not only to entertain elites but also to provide seeds and cuttings for imperial plantings. Of necessity, botanical gardens mimicked the ecologies of the places from which plants were taken; but they also packaged those environments, making them experimental ecological machines. Every colony gained its own botanical garden. Indeed, many exotic plants escaped from this site, sometimes becoming agents of multispecies invasion.

FIGURE 32. **Detail of *Empire* (3).**

FIGURE 33. **Detail of *Empire* (4).**

Given the extensive range of imperial terraforming projects, it is not surprising that feral effects abound. Plantations developed new kinds of weeds, pests, and parasites—such as witchweed, encouraged by the quick and dirty practices of international land grabs in Mozambique, and inherited by peasants who farm the land the grabbers have abandoned. Serena Stein's research has shown the importance of this parasitic plant, which wilts peasant crops on abandoned plantation lands.[34] Or consider the invasive lantana, escaped from colonial botanical gardens in India to take root in teak plantations, as described by Ursula Münster.[35] Lantana grows tall in those teak plantations, and it blocks the vision of the elephants that have survived the teak-harvesting work camps. With their vision blocked, elephants are taken by surprise by human visitors, and they sometimes react with deadly attacks. As Münster sums up the situation, "Lantana invades teak plantations and turns elephants violent."

Waterscapes as well as landscapes have been reshaped by imperial ecological experiments. Even the oceans are implicated. Imperial shipping has carried exotic species from one continent to another. Architect and historian Lionel Devlieger documents the travel of mitten crabs from Chinese harbors to Germany with the warships Germany sent to China in the nineteenth century.[36] The crabs now live in European rivers, where they accumulate heavy metals and other pollutants.

Meanwhile, even on the high seas, imperial rule matters for earth ecologies. The ships of great powers still rule the seas, and this rule, too, has feral effects. Sonic scientist Suzanna Blackwell studies the effects on whales of the blasting noises of gas exploration.[37] Whales communicate through sound, making use of the ability of water to carry sounds across long distances. But their ability to communicate is blocked by the noises of ships, especially the loud blasts of air guns used for gas exploration. Ship sound also travels over long distances. One study Blackwell cites identified air gun sounds 4,000 kilometers from the survey vessels that produced those sounds. This is a terrible thing for whales. Bowhead whale mothers guide their calves through sound; now this is drowned out by the noise of ships.

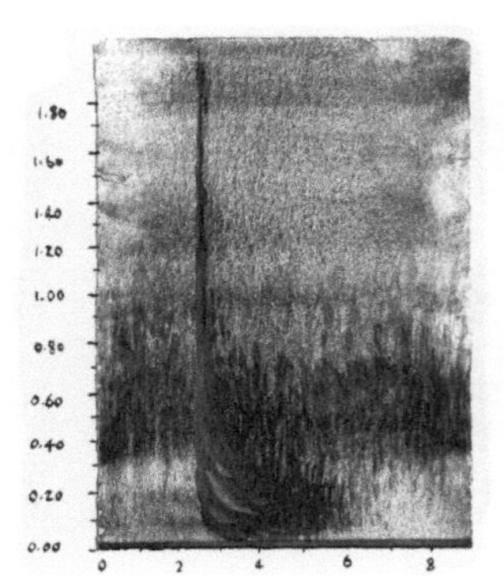

The machine-like modularity of imperial ecological experiments crosses the planet's surface, producing feral effects that haunt contemporary life, human and not human. Terraforming governance (or empire) is one of *Feral Atlas*'s four Anthropocene Detonators: invasion, empire, capital, and acceleration. We hope readers will learn more about these Detonators and the landscapes that help explain them, on the digital site, and that you will find the field reports of feral effects nesting in each. These infrastructure-building programs, we argue, respectively *detonate* the Anthropocene. They make even small patch disturbances significant at a global and planetary level.

FIGURE 34. *Alabama Fields.* 2013.
HELENE SCHMITZ

PART III
Histories

Anthropocene time is not just the time of rupture; it is also the intertwining of human and nonhuman histories through which more-than-human trajectories are made. This section explores new ways to appreciate the historical agency of nonhuman living beings as they intertwine with human histories, that is, on human timelines. For this kind of time horizon, a consideration of feral effects of human activity is useful; there is something particularly vivid about watching humans and nonhumans change together *in the same time periods*. The chapters in this section argue that feral biologies offer a privileged view of nonhuman beings as protagonists of more-than-human histories. Feral biologies show us nonhuman history within the events and timelines of human history.

Human history was classically understood in relation to the ways elite humans, such as generals and kings, changed their views and behaviors to address shifting challenges. Since the 1960s, social historians have added the stories of subaltern humans. Until quite recently, however, nonhumans have not been considered this kind of historical actor, that is, an actor capable of changing in relation to challenges. Academic legacies in both the natural sciences and the humanities have kept us imagining nonhumans as affecting history only by being present or absent, or more or less in quantity. Nonhuman nature, in other words, has been seen as essentially static. In both the natural sciences and the humanities, twentieth-century frames denied the historicity of nonhumans. In recent years, however, scholarly advances have made new kinds of history possible, including more-than-human histories.

The neo-Darwinian modern synthesis of twentieth-century biology deserves considerable blame for removing history from biological thought. Nineteenth-century evolutionary history was historical, exploring changing species' natures over time. In the twentieth-century modern synthesis, the idea of evolution was coupled with Mendelian genetics to show how organisms inherited the genetic nature of their parents. Slight genetic differences mark every individual, but at the level of the species, biologists believed, a static common heritage lasted until the species went extinct. Each species, static and ahistorical, gave rise to new generations, each of which lived and died on its own inherited merits.

All of this thinking changed at the end of the twentieth century, hoisted by its own petard. As the genetic heritage of each species became easier to describe, it became clear that organisms are evolving all the time, at least at the level of small genetic changes, many of them in response to changing environments—that is, historically.[1] Although many genetic elements of species remain untouched over long periods of time, the assumption of static heritage was wrong. Furthermore, genes may or may not be expressed—depending on histories; epigenetic expression can bring those histories into the next generation.[2] It also turns out that genetics can change through mechanisms other than mutation and parental inheritance. Bacteria, for example, are busy exchanging genetic material with one another all the time. Today, it's common to hear biologists speak of organisms "borrowing" genetic material from whole other kingdoms, even when the mechanics of how that borrowing happened are not yet completely clear. Such mechanisms depend on the fact that species turn out to have relations with other species that far exceed the prey–predator binary. In the new thinking, multispecies organisms might be considered playgrounds for bacteria, that is, sites in which bacterial relations develop a multitude of forms. Species are no longer considered autonomous units of evolution; instead, "holobionts"—that is, congeries of species living with one another—are units of evolution. For multispecies organisms, the holobiont includes the microbiome; with humans, for example, this involves the bacteria in our gut and on our skin. Interspecies encounters shape beings.

All of this brings life processes back into historical time. Organisms change in relation to the worlds in which they find themselves, whether through genetic changes, epigenetic expression, adaptation, or the company they keep, including their microbiomes. This makes them robust protagonists as subjects of history. History responds to their changing natures, even as they respond to history. There are other shifts in the natural sciences that lead in this direction—from disturbance ecology to geological process.[3] But the biology story is so vivid and thrilling that it does well standing in for the sea change. History is back, revealing the dynamic unfolding of shared pasts and futures in startling new ways.

The humanist story of the exclusion of nonhumans from history is older and perhaps more intransigent than that in biology. The problem stems from the Enlightenment commitment to Man as a special kind of being, making history through free will and consciousness. Since beings other than humans are thought not to have free will and consciousness, they must be, as Heidegger put it, "poor in world."[4] If beings cannot make worlds of language and meaning, or so it is claimed, they are only pawns and resources for human ends. In this legacy, nonhumans may be imagined as *reactive* to human action, but they are not themselves historical actors, able to create new human and nonhuman trajectories.

Against such sentiments, a vibrant set of movements in the humanities has endorsed the importance and lively activity of nonhumans. Environmental history, political ecology, and science studies have been key participants; feminist theorists, from Val Plumwood to Donna Haraway, pushed far past Heidegger.[5] In the last twenty years, the environmental humanities have become a respected field of scholarship. This *Field Guide* would not have been possible without this development. At the same time, the worry that attention to nonhumans might usurp the gains of the humanities lingers, a murmuring in hallways. Our argument in this section is that this anxiety is misplaced: Bringing nonhumans into humanist discussion can enrich the gains of humanism rather than destroying them.

The three chapters that make up this section, all written by Anna Tsing, show how more-than-human histories are composed—and how they matter to humans and nonhumans alike. Chapter 7 discusses the history of capitalism, following Eric Wolf, to show that nonhuman organisms are active historical agents. Chapter 8 opens up concepts of history further to develop a kind of historical storytelling in which landscape becomes a nexus for multiple historical ontologies—that is, ways of doing history, human and nonhuman. Chapter 9 tells horror stories showing what human infrastructures are doing in the creation of dangerous feral biologies. What if history—both despite and because of human efforts—is no longer ours to guide?

SEVEN

Others Without History

ANNA TSING

Plants, fungi, and bacteria make their own history, but they do not make it as they please.[1]

What histories matter in the making of the capitalist world system? When Eric Wolf asked this question in the 1980s, it was easy to confine his research to humans. *Europe and the People Without History* remains a towering work in the humanistic social sciences because it opened the door to subaltern and Indigenous histories in the making of the capitalist world.[2] However, Wolf's tome does little to explore nonhuman histories; nonhumans for him are resources for human livelihoods. Today, we might take another look at whose histories matter. By extending Wolf's analysis to include nonhumans, we can push forward the points made in Part II of this book about how infrastructures shape the Anthropocene. This allows me to make the following argument: Nonhuman living beings gain new "agilities" (that is, historically developed abilities) through their relations with the infrastructures of capitalism; in the process they become Anthropocene historical protagonists.[3]

This chapter discusses two plants and two animals, all masters of intervention: They are entangled in human affairs, but they lack the docility of domestic creatures or the malleability we imagine of "natural resources," waiting for human use. They are neither domestic nor wild, neither dependent upon us nor outside our world. Their histories are all mixed up with ours, but they are out of our control. They remake themselves in the conjunctures of our histories, and they change the possibilities for humans, and other creatures, to proceed. Because they force our hands, we can't ignore their activity. Because they behave so differently in varied situations and across time, we are pressed to acknowledge their transformations and the effects these have on our futures. Their in-your-face actions seem a good place to see cross-species historical trajectories in the making.

Why Articulation Matters in Telling Histories Today

But why use Wolf's form of analysis, when other kinds of "history" are possible? To tell nonhuman and human histories together, without reducing nonhumans to puppets of human dreams and schemes, requires attention to how disparate phenomena might link without fully subsuming each other. In social and cultural theory, this kind of work is done by theories of articulation, that is, non-necessary links. Articulation, as theorist Stuart Hall explains, is both linking and expressing; articulation ushers new social histories into being in the coming together of quite different trajectories.[4] Such theories have become less common in the social sciences and humanities in recent years as structural analyses have had less appeal. But just as articulation theory once helped scholars bring people of color in the Global North and people from the Global South into historicity, so too can this kind of theory build more-than-human history.[5] Wolf's work is useful in reconsidering the analytic uses of articulation.

Wolf, like a number of other anthropologists in the 1980s, imagined the history of capitalism as developing through a contingent series of articulations with other modes of production.[6] This approach disappeared in the 1990s with the flood of academic attention to globalization. Afterwards, capitalism (and with it, modernity) was increasingly analyzed as world engulfing, with historicity emerging from perturbations inside the beast.[7] As scholars posited the enclosure of capitalist modernity as a world without outsides, they lost the craft of attention to unfamiliar and uneven developments. Today's anthropological excitement about ontological difference derives in part from its response to that shutting off of attention.[8] But Wolf's solution should be equally intriguing for contemporary readers. In considering articulations across ways of life—human and nonhuman, and whether or not they are modes of production—we can attend to contingencies and conjunctures. At the historically productive edges of capitalist hegemony, capitalism is constantly remade.

Articulation offers a framework for attending to radical difference—but refusing to be stuck in its reification. Wolf showed us worlds that Western and non-Western people became involved in together, and often without common agreements or sympathies, that is, *in the midst of* difference. This works through the pleasure of Wolf's historical narratives, from the contingent encounters of the North American fur trade to the global movements of labor and commodities. The stories carry the framework. As we revive a critically reflexive empiricism, such arts of telling will be useful.

This chapter explores how these approaches might open the history of capitalism to nonhumans. I do not tackle capitalist struggles to tame organisms as resources, although I see that as a worthy topic.[9] Instead, I've chosen pests and weeds: organisms whose historicity is evident precisely because they are never fully integrated into capitalist designs. Haunting capitalism, they change its history, even as capitalist history changes them. They are not anti-capitalist heroes; they thrive with the infrastructure of capitalist modernity, even as they use that infrastructure to do unexpected things. The engineers of capitalist modernity inadvertently encourage them—and yet despise them, revealing their structural difference. By attending to articulations between the world-making projects of these organisms and capitalist world-making, I show the active, morally ambiguous, and feisty historicity of these nonhumans. Any good history of capitalist world-making needs them.

Engineering Fan Club

Water hyacinth is a floating plant with beautiful lavender-colored flowers. Before the late nineteenth century, it was found only in Amazonia, and perhaps adjacent regions of South and Central America. It grew in quiet backwaters of rivers and in lakes. It was not a troublesome plant. When rivers ran in flood, its floating mats were swept away by the fast water. This is still the case in much of its homeland. As long as the river is running strong, water hyacinth is unobtrusive. But engineers made a world of calm water, encouraging its increase.

In the late nineteenth century, water hyacinth became a trophy for the worldwide expansion of botanical gardens. European centers all wanted it; by 1883, it was in Göttingen, Berlin, Kew, Edinburgh, Glasgow, Monaco, and Strasbourg.[10] In 1884, it was featured at the Louisiana Cotton Centennial Exhibition in the United States. Japanese were said to be distributing it as souvenirs, so it must have already gone to Japan. The botanical garden in Java obtained some water hyacinth in 1894. What happened next is representative: "As the plant grew rapidly to unmanageable proportions, cartloads were thrown in river Tji Liwung crossing through the garden." Soon enough, the plant could be found across the anthropogenic landscape on "river banks, lakes, canals, railway ditches, morasses, pools, tanks, reservoirs, and also in paddy fields."[11]

By 1893, water hyacinth clogged St. John's River in Florida; by 1918 drainage canals had facilitated the plants' spread from that river to waterways across southern Florida.[12] In 1904, the botanical garden in Java gave a sample of the plant ("the gift of Java") to King Rama V of Thailand. By 1908, it had spread across the Chao Phraya delta and into the Mekong River, where it spread across French Indochina, just as colonial engineers were remaking the waterscapes.[13] Most of Europe was in luck: The plant did not thrive north of the 37th Parallel.[14] But the waterways of Africa, Asia, and Australia were increasingly choked by water hyacinth. South America too. How could it be that this once polite plant was suddenly smothering water bodies everywhere?

Appreciating the magnitude of the problem requires introducing the plant in its current, widely distributed form. Given a calm water surface and warm enough temperatures, this is a plant that knows how to fill the available space, becoming—and with astonishing speed—an unbroken single-species mat across the water. In Louisiana, one plant became 65,000 plants in a single spring season. In Mississippi, the water surface covered by the plant increased by 6–10 percent *every day*.[15] Once it fills the surface, the mat just gets denser and denser, until a person can walk across the water on it. Of course, this disrupts water transport—and even the flow of water.[16] In irrigation canals, water hyacinth reduces waterflow by 40–95 percent.[17] The plant hungrily sucks up fluids, drying out local water sources, such as village water tanks. It shades out other water plants and kills the fish and waterfowl that depend on them; fisheries are destroyed. Oxygen in the water is depleted, and bad smells develop. In rice paddies, water hyacinth interferes with seed germination and seedling establishment. It forms a habitat for rats and for the vectors of many human diseases—including malaria, encephalitis, schistosomiasis, filariasis, and more. Both villagers and elites suffer.

How could such an astonishing spread have been accomplished? Here the story turns to the intertwining of human and water hyacinth histories. Engineers made a new world of water; and water hyacinth responded by learning how to expand into it.

Water hyacinth followed projects of modern water management; it became the engineers' fan club. Dams created calm open reservoirs. Irrigation and drainage canals tame the water and often slow it down. These were the places to which water hyacinth migrated. After water hyacinth's introduction to a new area, it generally remained a quiet resident—until the waterscape was remade for it by engineers.

Water hyacinth was introduced to Kyushu, Japan, in the late nineteenth century, and a poet noted it in urban canals in the 1920s.[18] But this plant, which in some places takes over water bodies in a few months, did not become a menace until the 1960s—after flood control engineering seized the country following disastrous floods in 1953. In Egypt, the plant was introduced by a colonial administrator in the late nineteenth century but grew only in urban canals through the first half of the twentieth century. After the building of the Aswan High Dam, which slowed down the Nile River, the plant spread rapidly throughout the Nile Delta.[19] In the Mekong Delta, the spread of water hyacinth and the spread of colonial water engineering went hand in hand. And in its South American homeland, water hyacinth began causing problems—with engineered water. In one example, Suriname's Brokopondo Reservoir began filling in February 1964; by September, 9 percent of the reservoir was covered, and by April 1966, when removal began, 53 percent. Meanwhile, by 1982, 650 tons of water hyacinth were removed every year from the Panama Canal to prevent the waterway from clogging.[20]

Worse yet, water hyacinth flourishes with the provision of excess nutrients into the water. The runoff of fertilizers, sewage, or nutrient-rich factory waste supercharges water hyacinth, allowing it to outcompete other water plants. As one commentator wrote, "Irrigation, fertilizers, and flood control are essential parts of the green revolution, but the more successful these techniques are, the worse the weed problem becomes."[21]

Green revolution scientists, who treated all living things as if they were industrial resources, first hoped that water hyacinth could be defeated by importing its homeland predators. ("Predator escape" theories of weedy success fit with a world of static and essential beings affected only by competition and war.) But no: Water hyacinth has no special predators in Amazonia. Instead, water hyacinth has changed to meet its opportunities. Simone de Beauvoir's aphorism about women can be transposed to weeds: One is not born, but rather becomes, a weed. Weeds are formed in relations, and, through relations, they make interspecies history.

As will be explained in Chapter 8, the water hyacinth that follows the train of water engineers has lost its discrimination in mating partners. But there is more: Water hyacinth in diaspora often doesn't mate at all. It just extends itself clonally. Diasporic spread has selected for vegetative growth rather than sexual reproduction. Water hyacinth plants around the world are closely related genetically.[22] There is a reason they have not needed the fine-tuning to local environments that sexual reproduction allows. The environment is homogeneous. Water engineers have made it so. The worldwide clone of water hyacinth follows the worldwide clone of water engineering projects.

How are histories of the colonial proliferation of water engineering and the proliferation of water hyacinth *articulated* with each other? A return to Wolf can show the analytic moves involved. Wolf urges readers to consider consequential articulations between vernacular local forms, on the one hand, and those of European traders and capitalists, on the other. Capitalism often changes local ways of being: The Mundurucu of Brazil, for example, became manioc producers for the global trade and in the process changed their kinship and settlement patterns.[23] Such articulations were shaped not only by capitalist logics but also by Indigenous ways of being. Manioc entered the global trade precisely because Mundurucu women were so skilled in producing it; women's increased manioc production for the world market anchored newly matrilocal villages. A different example of articulation is more terrifying: Portuguese traders were able to obtain enslaved people in the Congo because Congolese kings initially facilitated the exchange of people for prestige goods.[24] The royal system, based on exchanges of women, dependents, and goods in lineages, disintegrated under the pressures of the slave trade; guns made lineage-based authorities irrelevant. But the trade would not have begun without the divergent social forms of Portuguese traders and Congolese kings, whose differences came together to produce something unfamiliar to each. The initial articulation mattered, even as it quickly fell apart.

The same kind of argument illuminates historical relations between humans and water hyacinth. An initial set of articulations spread water hyacinth around the world, as botanical gardens coveted its showy flowers. Soon enough, however, the articulation changed, just as it did in Munducuru kinship and settlement. Plants selected for vegetative growth and self-pollination proliferated against the will of the engineers, who made reservoirs and canals as if for those plants. Wind and water in anthropogenic waterways tapped the plant's free-floating habit to move it around. Water hyacinth took off, and botanical gardens became irrelevant. Water hyacinth reminds us that plants—molded in the conjunctures of history—develop their own trajectories and logics.

Property Spook

Capitalism as a system depends on stabilizing *things* as economic resources. First, if things are commodities, they must stay intact and under control at least for the period of their transactions. Second, to become wealth, things must stay put long enough to be collected, and later transacted again. There is a reason that Marx used a stretch of linen to show us the basic mechanics of capitalism.[25] Linen can go into a transaction and out again without too much damage, and it can be folded and stored in a warehouse to take part in those famous M-C-M transfers through which accumulation is enacted.[26] We can only define a property system by imagining things as essentially stable, like linen.

But, of course, lots of things are not stable at all. Much attention has gone into the problem of things that deteriorate and disappear; to make them property, we preserve them or sell them off as quickly as possible.[27] But what of things that move around and proliferate beyond control? Such things may not be good to live with, but they are good to think with. They force us to notice that property requires a host of chaperones; it is not an exclusive relation between a person and a thing.

To make this case concrete, consider the dilemma of homeowners in the United Kingdom. A house may be in good condition, but it is very hard to sell it if a particular garden plant can be detected nearby. Mortgage companies and insurance agents will shun the home's owners. The property may stop being property, in the sense of a transactionable commodity. The plant that thwarts such homeowners is Japanese knotweed, a well-known property spook. Once it is growing somewhere, it is really hard to get it to leave. And it just keeps growing, coming up through the basement, cracking concrete, shattering foundations, refusing to be eliminated.[28]

As capitalism continues to eliminate vulnerable species and landscapes, we are forced to live more and more in a land of weeds, aggressive species that both love our disturbances and thumb their noses, so to speak, at our conceits. Such species expose the lie that modernist problems will be solved with modernist solutions. By thwarting basic categories such as property, they show us a future in which even the simplest frames of our political economy will survive only in ruins.

Japanese knotweed made its debut in Europe with all the pomp of property. It was part of the private collection of Philipp Franz von Siebold, a German doctor who used his unusual access to Japan to collect plants, which he brought back to the Netherlands to offer on the European market.[29] This plant won a gold medal from the Society for Agriculture and Horticulture in Utrecht in 1847 for the most interesting new ornamental plant of the year; with this honor, it was offered at the exorbitant price of 500 francs for twenty-five clonal stems. Siebold's collection had only one plant, and this a polyploid female (which turned out to be consequential), but it was easy to produce vegetative clones for sale and sale-promoting distribution. For this last goal, Siebold sent some to Kew Gardens in 1850; Japanese knotweed quickly became a much-valued garden plant in Britain. It was promoted particularly in the "wild garden" movement, where it was used to create an informal look and shady woodland walks.[30] Here too one might see the triumph of property, as inflected by empire: Britain's informal gardens were claims to own the whole world, through exotic plants.[31]

By the early twentieth century, however, Japanese knotweed had earned a reputation for escape—and non-extractability. In the 1930s in East Cornwall, a new name for the plant maligned a local commercial nursery: "Hancock's curse."[32] Meanwhile, reports began to come in not just from northern Europe but from every place the European diaspora had spread the plant. Because it extends itself through rhizomes, it spreads through the countryside. Because it forms a dense shady cover, and secretes noxious chemicals into the soil, native plants and wildlife disappear around it.[33] Because it loosens the soil, riverbanks erode. Because it grows back readily from tiny scraps of rhizomes, cutting it down is useless as a way to remove it. Worse yet, removing the scraps and the soil and putting them somewhere else only spreads the plant to that new location. And then, too, there it is drilling into drainage pipes, roadbeds, and people's basements, destroying property.[34]

Botanists took a look at the plant—and, lo and behold, it wasn't just one thing. Its inability to be stable as property extended into an inability to be stable as species. Recall that Siebold had imported just one individual—a female. That female, in all its clones, produced copious flowers of great interest to bees and other insects, which brought pollen from all around. Surprise: There were many fertile matings. The original clone produced offspring with two other imported genus mates: a small knotweed from the Japanese mountains, and a giant knotweed from northern Japan and Sakhalin Island.[35] More surprisingly, viable seeds were also produced in matings with more distantly related ornamental plants such as Russian vine (*Fallopia baldschuanica*, in *Fallopia* section *Pleuropterus*, unlike knotweed in section *Reynoutria*) and even an Australian native in entirely different genus (*Muehlenbeckia australis*).[36] Commercial nurseries were the ideal place for such illegitimate sex, since all the exotic plants were gathered together; but such matings could also occur in those British gardens, proud to gather the world.

How vigorous the new hybrids were. Not only did their variety allow knotweeds to colonize new climatic zones, but they also selected for those that could easily regenerate from rhizome fragments, and in any kind of soil.[37] Through promiscuous matings, the plants experimented, finding just those variants that might be most successful in thwarting human attempts to curtail and control them.[38]

Property spook: It doesn't eliminate property, but it haunts it. Consider now an organism that had more decisive effects.

King Killer

In *Empire of Cotton*, historian Sven Beckert masterfully argues that plantations worked by enslaved people ("war capitalism") were the necessary complement to factory production ("industrial capitalism").[39] At the center of the synergy between these two forms is the history of cotton. Cotton, he argues, led the way for capitalist history more generally. The heart of Beckert's book is the nineteenth century, when these two forms worked together to create a revolution. But this book is not just about time; there are also *places* here: the mills of Manchester and the plantations of the U.S. South, where King Cotton ruled. At the end of the century, however, the centrality of these places disappeared as cotton production spread across the world. What happened to King Cotton? A lot, including the disruption of the U.S. Civil War. Still, after the war, cotton production was resumed. It was only during this resumption that the hegemony of the U.S. South as the epicenter of global cotton production was finally dismantled. Yet southern cotton's chief antagonist from that time hardly figures in Beckert's history. It was an insect: the boll weevil.

The boll weevil crossed the border from Mexico to the United States in 1892, in Texas; by the early 1920s, its range covered the entire U.S. cotton belt, where it caused enormous devastation. Scientific farm management, born from this moment, urged labor-intensive interventions. Meanwhile, white violence against African Americans increased as white landowners panicked.[40] Black tenant farmers and sharecroppers deserted cotton growing in droves. Cotton planters tried innovations, but these raised the cost of production even on those rare occasions when they worked. Scholars disagree about just how momentous the effect of the boll weevil was on the turn-of-the-century South, but one thing is clear: The South could no longer set the terms for global cotton production.[41] Cotton production had begun elsewhere during the U.S. Civil War, which had disrupted the South's hold on the trade; now it took off in Brazil, Egypt, India, China—and more. The king was dead.

It is hard to think of a more vivid case of an organism that changes the trajectory of capital. Yet the boll weevil brings to mind other crop antagonists similarly stimulated by the ecological simplifications of the plantation. For example, banana production in Honduras as described by historian John Soluri was continually disrupted by fungi.[42] Every time the fungi became strong, the owners abandoned the place and moved their operation elsewhere; thus the banana belt took up an enormous swath of land, most of it in ruins. This is surely consequential for capital. Still, even the environmentally sensitive historian Soluri fails to make the fungi protagonists; they form the background against which owners and workers struggle. Admittedly, it is hard to capture our imagination for rust and rots as protagonists. In contrast, singers and storytellers have done the heavy lifting in making the boll weevil a consequential character. Here is blues guitar legend Lead Belly:[43]

> The first time I seen a boll weevil,
> He was sitting on a square,[44]
> Next time I seen a boll weevil,
> He had his whole family there,
> He's a-looking for a home,

The boll weevil emerges in song as a person, a history maker. Sometimes he is tested, but he comes out victorious:[45]

> The farmer take the boll weevil,
> And he put him on the [red hot] sand,
> The boll weevil, he says to the farmer,
> "You are treating me just like a man.
> I'll have a home."

Key to the story, too, is Lead Belly's identification with the boll weevil, through his own displacements. While it is surely a simplification to attribute everything to one insect, it has been convincingly argued that the Great Migration of African Americans out of the rural South to industrial jobs in the North was sparked and abetted by the boll weevil. Former cotton workers were looking for a home:

> If anybody should come along and ask you people
> Who composed this song?
> Tell them that Huddie Ledbetter.
> He's just been here and gone

He's a looking for a home
He's looking for a home.

Lead Belly has made the boll weevil a protagonist of the same status as himself.

The cotton boll weevil belongs to a group of pollen-eating insects that specialize in plants of the mallow family, which includes cotton.[46] Our protagonist seems to have come to specialize in cotton in the place it was domesticated in Mexico.[47] Still, as long as cotton is grown in small patches and mixed with other plants, the boll weevil is just one of many annoying pests. It took the plantation to bring it to full epidemic status.

In the autumn, after the good eating is done, boll weevils disperse from their cotton plants, and even though they are not good fliers, some travel more than 100 kilometers, carried by the wind.[48] They smell cotton and can crawl to it, but they have little control over their flight.[49] It takes a lot of weevils and a lot of cotton to bring the regional population to pest status. This is where plantations come in handy, from the weevils' perspective. Concentrate all the cotton in one place, and the weevils will eventually drop into it.

While the boll weevil likes cotton, it is willing to eat from many plants, from many different families.[50] That dietary range—as well as the habit of metabolic resting when the weather is cold—keeps it going when there is no cotton around. There is one thing, however, that all those other plants cannot do for it: give it offspring. Only cotton makes the insects sexually and reproductively active. Then they are quite active, each female producing more than two hundred eggs, exponentially multiplied by eight to ten generations in a season. Every larva ruins one cotton bud, a potential boll.

The boll weevil has other tricks, too, that help it settle in new places. The time it takes to develop from egg to adult and the time it spends in metabolic rest can vary strikingly: This plasticity is useful in adjusting to new climates.[51] Unlike the other organisms described in this chapter, it did not have to change its reproductive apparatus or hybridize with others to become a major pest. All it took was the plantation, which concentrated its reproductive abilities. A cotton plantation is not just a home for the boll weevil: It is a singles party, an apartment building, a maternity ward, and a warehouse full of fertility-granting food.

Indeed, after deposing King Cotton in the U.S. South, the boll weevil kept moving. In 1983, it reached the cotton fields of Brazil and less than five years later covered 90 percent of the cotton-producing area. In 1996, it entered Bolivia. In 2006, it reached the center of Argentine industrial cotton.[52] There farmers with more than 900 hectares produce 75 percent of the nation's cotton.[53] The overthrow of kings continues.

Game Changer

People today are surrounded by a vast machinery working to naturalize capitalism, property, and modernist landscape engineering; we are led to believe there is no alternative. One reason anthropologists write about capitalism is to show our readers that other arrangements are possible. Anthropologists draw inspiration here from allies outside the academy, including farmers. Sometimes other creatures help us. These are not the darlings of green revolution science, the biological controls. No, these are the pests themselves, within the right conjunctures, that mobilize more-than-human communities of action. The story here turns then from a pest that stopped capitalist trajectories to one that stimulated an alternative.

The golden apple snail was first brought from Argentina to Taiwan as part of the kind of sure-to-fail development project do-gooders have so often helped to promote.[54] The snail was supposed to be a protein source for rural people. Not surprisingly, no one would eat it. The snail, however, liked very much to eat. A creature of water, it found itself quite at home in paddy fields, where it proceeded to consume the rice seedlings—and everything else it could find. In the village my colleagues are studying, it was deliberately introduced to eat water hyacinth; instead, it eats rice.[55] From Taiwan, it was transferred across Asia on similar ridiculous premises; everywhere it eats rice, while also reducing indigenous wetland ecologies to barrens.

A poison that can kill a snail is called a molluscicide. Agroindustrial policy in Taiwan, shaped since the 1950s by the U.S.-led green revolution, has promoted molluscicides. The problem is that a poison that can kill snails—and especially numerous large snails scattered throughout the waterways—will also kill you and me. Furthermore, farmers have noticed that a snail hit by poison uses its remaining time on earth to lay as many eggs as possible, thus assuring an abundant next generation. The poison kills us—and increases the number of snails.

Rather surprisingly, this situation mobilized a group of young people who were already dedicated to revitalizing Taiwan's countryside but still casting around about just how to do so. Confronting the snail, they found themselves. In communities dedicated to what they call "friendly farming," they refuse molluscicides, instead laboriously handpicking snails in the middle of the night when the snails are most active. Meanwhile, they compose songs and tell stories about snails; they have instituted local festivals in the snails' honor. "Snails unite us," they say. They use the snail as an icon to develop their political platform.[56] They don't try to exterminate the snail. Each year, after the rice is sufficiently mature to withstand the snails' onslaught, these farmers stop removing snails, instead enrolling the snails as allies in weed removal. They wait until the following spring to resume snail harvesting—and then work like crazy to outrun snail reproduction.

Seasonal bursts of labor-intensive handpicking? This is hardly a modernist solution. For the friendly farmers, it is part of a bigger program. In the fall of 2016, before the plan was unilaterally ended by the United States, the young farmers were idealistically butting heads with the Trans-Pacific Partnership. They imagined an interruption in business as usual: a chance for environmentally friendly small farmers. In strange alliance with the snail, they argue that another world is possible.

Living in the Capitalocene

While it is a pleasure to describe emerging alternatives to capitalist agribusiness, most of the time, capitalist trajectories don't make way for alternatives, instead rolling right over them. Capitalist modernization, despite so many criticisms, has proved resilient. Even as progress stops making sense for so many people across the world, capitalist modernization rolls on. Horror stories of ensuing plagues are picked up in Chapter 9, which shows how capitalist practices empower feral pathogens, sometimes ending lifeworlds. Still, admitting to this force is no reason to imagine capitalism as a homogeneous, enclosed world. Articulation theory remains important to notice continuing, and renewing, zones of difference.

Sociologist Jason Moore has written an insightful book about the ecological dynamics of capitalism, a "world ecology" he calls the Capitalocene.[57] The book usefully argues that "cheap nature" has been as important a product as "cheap labor"; capitalism has remade landscapes for its best use. The book refreshes discussions of ecology and capitalism. However, Moore's imperative to get beyond the false dichotomy of Nature and Society drives him to posit an all-encompassing capitalist system for which there are no outsides. For Moore, nonhuman (and human) lives are always an expression of capitalist logics. Here nonhumans, once again, become mechanical puppets of human elites.

There are analytic alternatives that do not depend on reestablishing the dichotomy between humans and nonhumans. J. K. Gibson-Graham (who rose to prominence through their critique of monolithic anti-capitalist narratives as heroic masculine conceits) offer a different way to reach out to nonhumans.[58] Writing with Ethan Miller, they argue that the concept of (political) economy need not be limited to humans.[59] Nonhuman livelihoods might be considered part of the analytic domain "economy" along with those of humans. Capitalism controls some nonhuman lives more than others. Even the most controlled, too, mobilize livelihood processes that capitalism cannot organize for them.

Whenever capitalists attempt to capture value from processes outside capitalist control, such as photosynthesis or digestion, they encounter the dynamics of articulation. Beginning with this fresh start, I imagine capitalism *only* in articulation with other modes of production. Of course, this approach is an analytic choice, not just a description. But it matters to those who care about more-than-human livability. Articulation allows us to tell stories in which nonhuman social dynamics form better or worse accommodations with capitalist infrastructures. From this, one can see capitalist histories in the making. Some patches are more livable than others; this assessment allows us to pick sites of struggle.

The creatures discussed in this chapter are not compliant subjects of human histories; nor are they the heroic resistance. The kinds of feral beings described here are "camp followers" of capitalist modernization.[60] Like the washerwomen, suppliers, and service workers that once followed armies, they tag along with capitalist developments. But they are not trained to obey orders. They are an unruly rabble, and it is never clear when they will get in the way of generals' plans. They refuse to disappear when they're not needed. They force changes in the army's path, as when boll weevils took out southern cotton. They leave marks, from dying woodlands to choked waterways. "Notice us," they seem to say. If scholars want to understand the effects of the Capitalocene, we will need to notice such beings as they join us—for better or for worse—in making history.

There are other ways of doing history. Some are introduced in the next chapter, which also extends the discussion of water hyacinth as the engineers' fan club.

EIGHT

What Is History?

ANNA TSING

What kind of history can a plant make? Some plants, such as the diasporic water hyacinth introduced in the previous chapter, are so forceful and uncooperative with other species that it is hard to ignore their world-changing activities. But if we allow plants to be historical agents, what would history mean? This chapter takes up this question through revisiting some of the debates about history that have enlivened the field of anthropology—and then following some further adventures of that annoyingly prolific plant, the water hyacinth. The chapter traverses discussions of ontologies—that is, ways of being—that have vitalized anthropology in order to further reorientate our sense of history and historical agency.[1] Staying within the terms of that discussion, the chapter suggests that landscapes can be studied as ontological coordinations and miscoordinations in which both humans and nonhumans take part. This is a version of articulation theory in which varied elements of landscapes coordinate to create effects. Histories are made in such coordinations and miscoordinations.

Diasporic water hyacinth comes back in this chapter as protagonist because the plant is a home-wrecker, and not just for humans but for many denizens of freshwater ecologies. Its feral biology alerts us to how out of control our imperial and industrial engineering has become. When one considers these feral biologies, imagining that new imperial engineering will solve the crisis of existing engineering makes no sense.

As discussed in the Introduction, noticing is a small contribution to limiting the scope and range of the Anthropocene.

Recapitulating, water hyacinth (*Eichhornia crassipes*) is a floating plant with showy lavender flowers that originally came from the Amazon region; it was spread around the world in the late nineteenth century with the craze for botanical collecting. This was a high time, too, for colonial water engineering: dams, canals, reservoirs, river straightening, dredging, irrigation, embankments. Without human intent, water hyacinth proliferated together with water engineering; water hyacinth became a flourishing, floating diaspora. The plant quickly overwhelmed engineered water habitats, covering the surface of canals, reservoirs, paddy fields, water tanks, and the slow rivers that resulted from dams. As water oxygen levels dropped under dead and dying mats of vegetation, other freshwater organisms died. People suffered as well—not only because their transportation by water was blocked, but also because other uses of water—from fishing to irrigation to drinking—became impossible. The Bengal Delta, according to historian Iftekhar Iqbal, changed from the nineteenth-century rice bowl of South Asia ("Golden Bengal") to a twentieth-century morass of poverty, in part because so much cultivated paddy land and potential paddy land was lost to water hyacinth.[2] In 1943, 3 million people died in a famine there. Iqbal argues that water hyacinth played an important role in the impoverishment that made the region so vulnerable to inequalities and perturbations.

Some authors have taken to glorifying feral life for its rambunctious vitality.[3] But it makes little sense to glorify water hyacinth; the plant can be a killer. To say that plants are historical actors is not to approve of their actions: We need to break the habit of assuming that agency, action, and transformation are always good. But this is just an opening to the argument here. In what kind of history do plants participate? In the next section, this question is explored through debates in a particular field: anthropology.

Social Anthropology: Past and Present

Questions of the meaning and importance of history have animated the discipline of anthropology for most of its existence. One place to begin is E. E. Evans-Pritchard's Marett Lecture of 1950, "Social Anthropology: Past and Present," which opened debate in the influential *Journal of the Royal Anthropological Institute.* Anthropology, Evans-Pritchard argued, should be a historical science, a search for patterns rather than generalizable laws. History, he noted, had been discarded in the rejection of Victorian evolutionary schemes. "But with the bath water of presumptive history the functionalists have also thrown out the baby of valid history."[4]

Evans-Pritchard's thoughts on the relation between anthropology and natural science are especially interesting for a discussion of nonhuman historiography. Evans-Pritchard argued against a functionalism that apes natural sciences in the search for timeless natural laws. "It is easy to define the aim of social anthropology to be the establishment of sociological laws, but nothing even remotely resembling a law of the natural sciences has yet been adduced."[5] The search for patterns, he suggested, does not require the assumption of law-like normativity. Contrasting anthropology as history to the false scholasticism gained by law-seeking, he offered the following:

> Regarded as a special kind of historiography . . . social anthropology is released from these essentially philosophical dogmas and given the opportunity, though it may seem paradoxical to say so, to be really empirical, and, in the true sense of the word, scientific.[6]

Is this relevant for today's challenges? This chapter argues that the conditions for the contrast between history and science have turned over. The philosophical dogma from which (more-than-human) history might release us is the human exceptionalism of humanists. Historically minded natural scientists might become allies.

Evans-Pritchard could take for granted a natural science in which history was ignored in the search for universal laws; this was the dogma he rejected. Today, however, as explained in the Introduction to Part III, new forms of biology and ecology have become historical. Anthropology can find allies for historical analysis with important wings of biology and ecology. Together, we might ask what kinds of histories humans and other living beings kick up in common.

Evans-Pritchard's lecture opened a continuing conversation in British social anthropology. By the 1950s, however, the U.S. social sciences were beginning to displace British scholarship in claiming scholarly hegemony. It took U.S. historians' reinvention of social history in the 1960s and 1970s to make anthropology notice history in the United States and the places it influenced.[7] This new social history recovered records of forgotten people—peasants, rebels, women, queers—and it had direct appeal for anthropologists. By the 1980s, the new social history was making a big impact in American anthropology.

One important book in reorienting the field was Eric Wolf's *Europe and the People Without History*, discussed in Chapter 7.[8] Wolf's criticism of anthropology was that it kept "societies" analytically apart, ignoring global connections. But to make this point, he had to revitalize "history" to allow more kinds of people and events inside it.

Wolf's analysis worked through written sources in European languages, not the oral histories, material histories, music histories, and other alternatives that informed other parts of the social history movement. But social historians of that period, whether they stuck with conventional records or added alternative sources, tended to agree: It was possible to construct alternative histories *even using conventional archival sources*. During Evans-Pritchard's time, many anthropologists worried that without Indigenous written records, no history was possible. They thought the reports of colonial officials, missionaries, travelers, and even professional historians were too contaminated with Western biases to use them to construct anthropological histories. In contrast, Wolf embraced those contaminated sources alongside records gathered by oral historians and professional ethnographers. He made a history of "people without history" by drawing events from those records but reorienting them to his analysis, which stressed the articulation of modes of production in the making of global capitalism. In effect, he argued that anthropologists can read *against the grain*, that is, use contaminated sources to learn more than their original writers intended.

A recent example from a historian offers a vivid example of what this kind of reading can do. In *Slavery's Exiles: The Story of the American Maroons*, Sylviane Diouf combs through records to find traces of escaped enslaved Africans in the pre–Civil War U.S. South.[9] There are no firsthand accounts. The exciting thing about the book is that, although there are very few records, she finds a way to tell a rich story. She uses documents intended to keep slavery in place, such as legal and commercial transactions, but gleans another story. This approach raced through the anthropology of the 1980s and 1990s and continues today.

Wolf was not the originator of reading against the grain, but he showed its power to anthropologists. At the same time, this productive move opened a great debate that is still active—and one that is important for writing more-than-human histories, since nonhumans do not pen their own archives. Returning to the earlier thoughts from Evans-Pritchard's era, might the approach above be misguided—because history itself is an imperial scheme?

History as Culture and Ontology

The debate began almost immediately with the publication of Wolf's book, when Michael Taussig picked a dramatic quarrel with *Europe and the People Without History*.[10] Taussig called Wolf's book, among other bad things, "authoritarian realism."[11] Lumping the book with Sidney Mintz's *Sweetness and Power*, Taussig argued that Wolf naturalized commodities, reinforcing the spell of capitalism's commodity fetishism.[12] Taussig was condemning Wolf's history as well. According to Taussig, a history of connections made through commodities not only naturalizes commodities but also naturalizes history as a taken-for-granted totality. For Taussig, instead, history should be the practice of throwing up disconnected scenes each so disturbing that we cannot but be dislodged from our misleading common sense.

In throwing up Taussig's article—in the very spirit of disruption Taussig intended—this chapter provides a disturbing haunt for contemporary discussions of social history. Let's turn directly to today's ontological turn, and its most charismatic and clear-thinking spokesperson: Eduardo Viveiros de Castro.[13] If Viveiros de Castro does not like social history, it is because it is just another exemplar of Western colonial technologies. History is a modernist temporality, another way to silence the native. But the native will not be silent. Amerindian perspectivism, according to Viveiros de Castro, allows its subjects to cannibalize and digest history, turning colonial temporality into Amerindian cosmology. "Viewed from indigenous Amazonia," he writes, "history is only a version of myth."[14] History digested becomes ahistorical myth; myth decolonizes history.

Taussig's history is not excluded. Indeed, one might see it at the center of the maelstrom. Taussig inspired a Latin American anthropology centered on the stories of mestizos, who carry the violence of the conquest through their own hybrid identities. Taussig disavows the possibility of an authentic Indian voice; anthropology is only for mixed-up survivors.[15] The excitement of Viveiros de Castro's work today comes from his refusal of this Taussigian decree: Viveiros de Castro works to bring back an Indigenous voice. Amerindians, Viveiros de Castro tells us, continue in the strength of their mythologies, which digest and dispose of historical hybridities, including those provided by Taussig. If the whole enterprise of thinking historically is shot through with modernist ontologies, it's not enough to radicalize history.

Marisol de la Cadena shares Viveiros de Castro's goal of bringing back Indigenous perspectives from the shadow of their erasures, but her version is less stark. In her book *Earth Beings*, she finds that Indigenous perspectives in the Andes are also ahistorical.[16] Despite this, she pieces together a narrative of generational and event-driven change over time. This is history, but it is contaminated with the ahistorical, the view from Indigenous cosmology. It is history, but, in her phrase, "not only." History, but not only: Might there be multiple ontological projects stuffed within historical storytelling?

This question bounces this story back to the 1980s, when there was considerable excitement among anthropologists about alternative modes of historical storytelling. Richard Price's *First-Time* pairs the oral history tradition of Suriname Maroons with that told by colonial historians.[17] Sometimes the narratives hover around the same events, and sometimes they diverge. Price offers one at the top of the page, and the other at the bottom; he suggests *layers* of history made of interpenetrating but distinct narrative modes. In his later *Alabi's World*, Price offers three layers: Saramaka oral history, colonial records, and the accounts of the Moravian missionaries sent to convert the Maroons.[18] In contrast to Viveiros de Castro and de la Cadena's condensation of the West as unitary, Price's layers separate two distinct European voices: the colonial officials and the missionaries. Each offers its own frame, its distinct historical ontology. History—Western and otherwise—becomes multiple.

A related project of adding vernacular and Indigenous historiographies to what counts as history can be seen in Renato Rosaldo's *Ilon-*

got Headhunting.[19] The Ilongot do not have the formalized oral canon of Price's Saramaka, but they do tell distinctive stories about the past. Most striking is the relation between Ilongot modes of telling pasts and landscape. Places hold stories of the past: not only of those who lived there but also of those who touched and traveled through, both humans and animals. Places materialize Ilongot histories—whether of hunting prowess, of communal movements, or of World War II. This is worth considering for telling more-than-human histories: landscape is an important clue. If history is multiple, its multiplicity might be understood in relation to such alternative materializations.

Materializations make histories—and not just in landscapes. Indeed, between the 1980s and the present, this insight was followed up in investigations of historical archives in all their materiality—that is, as readers, shelves, books, and maps. Colonial archives have proved exceptionally rich as sites for anthropological investigation. Here the writings of Ann Stoler and Michel-Rolph Trouillot have been crucial. Trouillot showed how archives-in-action can work to erase certain stories, creating silences even where there are records.[20] In complement, Stoler looked at what archives produce: the categories through which we know not just the past but also the present.[21] She contrasts her way of reading with the earlier social history. Whereas social historians read archives against the grain, pulling out stories beyond those the archives were created to tell, she reads *along the archival grain*, attending to the cosmopolitics the archives work to build. Reading along the grain allows her to see a "historical ontology."[22]

In her work on colonial archives, Stoler explores one particular historical ontology, with its categories of race, gender, and governance. Might this technique be extended to see multiple ontologies within history? Sticking close to materialities, might we read World War II simultaneously along the grain of Ilongot landscape stories *and* along the grain of Philippine colonial archives? Might multiple acts of reading along the grain, indeed, be a better way to return to social historian's reading against the grain—and even Evans-Pritchard's search for patterns? Just as Annemarie Mol suggests that the body is multiple, might history too consist of multiple interrupting and entangled historical ontologies?[23] To explore this, it is time to move back, at last, to water hyacinth, in all its materiality.

Water Hyacinth Infrastructures

In its Amazonian homeland, water hyacinth grows at the edges of streams and in lakes. Annual floods sweep the floating mats away, and seeds embedded in the mud restart the population.[24] As long as water runs strongly, water hyacinth is not a pest. It is only slow or stagnant water that creates a safe haven for water hyacinth proliferation, especially in diaspora, where the right pollinating insects cannot be guaranteed. As discussed in Chapter 7, it was colonial water engineering that created those safe havens around the Global South. Canals, reservoirs, dam-slowed rivers, irrigation ditches, water tanks: These are places for water hyacinth. The story water hyacinth tells through its proliferation is a story of anthropogenic landscapes made through colonial water engineering.

What is meant here by landscape? Anthropologists were stuck for a few decades in a cultural geography discussion in which "landscape" referred to the view from a painting, a perspective that established the painter's power over the scene.[25] In contrast, "landscape" here refers to the sedimentation of human and nonhuman activity, which, taken together, creates places. Landscape is a busy intersection of contemporary action entangled with the traces of previous action. Because it offers signs of the past, landscape is amenable to archival practices. Landscape is not just an archive, but one can do archival readings on a landscape.

Most of what's written about water hyacinth is by agronomists and biologists; it doesn't tell social scientists much about the social dynamics of landscape. Water hyacinth's dependence on imperial water engineering is nowhere articulated but everywhere taken for granted. Much of the literature is generated by employees of that great engineering machine who are trying to figure out what to do about water hyacinth; the one thing they can never suggest is getting rid of the engineering projects that cause the problem. It's a frustrating, claustrophobic literature, even as it is full of thrilling details. For telling more-than-human history, it is wonderful to find two anthropologists and a historian—Ashley Carse, Atsuro Morita, and Iftekhar Iqbal—who describe water engineering landscapes. These researchers don't care about water hyacinth per se; they bring it up to be true to their accounts. Their stories are focused on infrastructure—and, through them, one can see how thinking through infrastructure might take the project forward in finding multiple, materially expressed historical ontologies. The next sections follow them—to the Panama Canal, to Thailand's Chao Phraya delta, and to colonial Bengal.

Ashley Carse's study of the Panama Canal zone, *Beyond the Big Ditch*, has been important in building the anthropology of infrastructure.[26] His account traces "infrastructural work," that is, what it takes to make and maintain infrastructure. This is infrastructure in a material sense; he is interested in canals and roads. Indeed, the excitement of the study is that he notices *both* canals and roads. Each is embedded in a distinct infrastructural project. The canal was built at the behest of U.S.-led international shipping interests. Roads, in contrast, were built as part of a process of national colonization of the countryside. Canals were made to bring the big ships through. Roads were made to facilitate the spread of farming, a national "conquest of the jungle" project of opening the countryside for colonists. These infrastructural projects have sometimes worked together, but they have also interrupted each other. Shipping engineers fear the erosion caused by road building and farming. Farmers find their options limited by the power of the canal. The two sides fight over what canal administrators call watershed forest but colonists call agricultural fallow. The landscape is pulled two ways. Watershed forests are part of the canal infrastructure because without them the canal's reservoir silts up; fallow is part of the road infrastructure because it is the destination of farmers.

Water hyacinth comes into the story as both a clog in the works and another example of interference between the two infrastructural projects. Neither canal people nor road people like water hyacinth, but it affects them differently. As mentioned earlier, water hyacinth is a Latin American plant, but it only became troublesome with the spread of stagnant surfaces. The Panama Canal, and the reservoir built to support it, Lake Gatun, were perfect places for the proliferation of water hyacinth. Engineers estimate that the canal would close in three to five years if water hyacinth were not constantly cleared.[27] A huge amount of work is required. Much clearance is done through mechanical equipment, but poisons such as arsenic have also been important. Carse shows the particularity of the canal administration when he quotes a public health officer: "From a sanitary point of view the use of arsenic in drinking water is to be condemned." Yet the decision is made to spray the water hyacinth with arsenic regardless.[28]

Colonists enter this story through questions of health as well as transportation. In the building of the canal, communities were resettled, becoming dependent on the new infrastructures of the canal zone, especially Lake Gatun. Earlier roads were flooded out. If the lake and its feeder rivers became entirely choked with water hyacinth, it would be useless for transportation. Carse gets involved through canal administrators' discussions about how much responsibility they have to clear the water for other canal zone residents. Water hyacinth, an unwanted child of the canal complex, interferes too with the road complex, drawing attention to both the entanglement of the two projects and their respective difference.

Infrastructures, Carse shows, don't all work the same way, even if they are each, if distinctively, tied to a cosmopolitan colonial modernity. Both of these infrastructures are colonial technologies, yet they have different frames for making subjects and objects. Carse makes it possible to consider how infrastructural work might be a way of parsing landscape for its multiple historical ontologies. Canals versus roads: Infrastructures are a way to explore multiple archives in the landscape. Piled on top of each other, they are layers, worth separating analytically, but in practice they are entangled.

Analytically separable but practically entangled layers: Andrew Mathews has called this "throughscape," that is, landscape understood as divergent complexes of material practice that weave *through* each other.[29] Note that a throughscape is a *pattern*—and looking for pattern, rather than law, was Evans-Pritchard's key insight about history.[30] Throughscape is a pattern formed through articulation, a piece of a theory of how more-than-human histories are made. Whereas Chapter 7 showed how articulations between human and nonhuman trajectories make history, the consideration of throughscapes shows how varied landscape elements articulate, making landscape effects as history.

How shall we see the story from the perspective of water hyacinth? Carse hints at world-making by the plant but does not get into it. But water hyacinth also creates an infrastructure—a third infrastructure for the canal zone. This would be a "multispecies infrastructure," a term used by Atsuro Morita.[31] To explore it further, Morita brings the discussion to the Chao Phraya delta—and its water hyacinth.

Morita uses the term "multispecies infrastructure" not to describe water hyacinth, but rather to discuss floating rice, a deepwater rice that responds quickly to water-level changes by growing long stems, up to several meters. Morita argues that the cultivation of floating rice should be considered an infrastructure even though a key element in the infrastructure is a living being. Morita, indeed, cites Carse's point that "watershed forest [is] *part of the canal infrastructure*."[32] Living beings can play a part in infrastructural work. Floating rice, like watersheds in Panama, are helpful to humans organizing water infrastructure; deepwater rice fields organize the flow of water in concert with irrigation and canals. Morita sees this as plants working together with humans. He implies that both humans and plants are served.

Once one allows living things in the door, might the notion of infrastructure also expand beyond human goals? Might other forms of ecological engineering—nonhuman as well as human—be considered infrastructure? The floating mats of water hyacinth are an architecture, impeding the flow of water and sucking it up voraciously. The mat encourages some life forms, such as mosquito larvae, and discourages others. If infrastructures are sociotechnical structures "essential to enable, sustain, or enhance societal living conditions," as one definition has it, must all of them be human?[33] Water hyacinth constructs a niche, a form of ecological engineering in which slowed water, low oxygen, nutrient uptake, and underwater shade benefit some water-living practices and not others.

If floating rice creates an infrastructure, why wouldn't water hyacinth? The difference is important. Floating rice empowers human-made landscape plans. Water hyacinth as infrastructure works against those plans. But might that make the expansion of the notion of infrastructural work even more intriguing? Nonhuman infrastructures are not human aids; those made by feral beings can show us how ferality jumps into human histories.

Like Carse, Morita is interested in layered, multiple infrastructures in the Chao Phraya's canal system. In any delta, he points out, terrestrial and aquatic infrastructures compete with each other.[34] Furthermore, each may be multiple. In a paper with Casper Bruun Jensen, he contrasts the canals built by Siamese kings—which asserted royal authority by making Bangkok the center of a hub—with the canals built by the Dutch engineer who came to modernize Thailand, which rationalized drainage and shipping according to European standards.[35] Morita and Jensen introduce the word "palimpsest" to show off these multiple infrastructures. This is another version of the throughscape: layers that also interact. Morita and Jensen are interested in the ontological frameworks of each of these infrastructures. Each reaches, they argue, to a separate cosmopolitics of land and water.

Morita discovers water hyacinth as a surprise.[36] It keeps engineers, and the anthropologists who study them, on their toes, he suggests. Might this surprise also be a sign of ontological gaps—that is, the rifts that maintain equivocation, keeping infrastructural projects from collapsing into a single unified plan?[37] If observers don't expect water hyacinth, isn't it because we don't grasp its world-making agilities, at least in the building of infrastructure? Water hyacinth offers an additional infrastructure to the canal system, one that exists in tension with both modern water engineering and the royal canals that continue to structure the landscape. It cannot be absorbed into either, although it is entirely dependent on their stretches of flat water. The throughscape is constituted in these interacting but disparate infrastructures. They offer distinct material "archives" for a history that is "multiple" in Annemarie Mol's sense of the term: more than one but less than many.[38]

Iftekhar Iqbal's *The Bengal Delta* is an important source for considering how infrastructures divide and intersect in making landscapes and histories. Iqbal also has two infrastructures in palimpsest: railways and waterways.[39] His focus is on how each impacts the other; he gives us a detailed look at both their interference and their inter-determination in landscape-making in the period between 1840 and 1943.

The story forms in the delta of the Ganges, within which new land is continually emerging at the intersection between sediment washed down from the Himalayas and sand swept in with the tides. The force of the river flushes the ingress of seawater, resulting in fertile lands for growing things. Farmers took advantage of the influx of new nutrients in floods. New land emerging from the water meant room for new farmers—and the inability of landlords to block access to land. The sediments supported multiple kinds of rice, from short-season and short-statured rices to tall deepwater rices that withstood floods. Immigrants poured in; trade prospered. All this made the wealth of Golden Bengal in the nineteenth century. Flowing water was an infrastructure supporting agriculture, immigration, and trade.

The infrastructure of flowing water was distorted by the building of railroads across Bengal, the first of which opened in 1862. This was the beginning of the global railroad boom: Excitement was high; investment funds poured in from London. Engineers were ecstatic. The Bengal delta was completely flat; no switchbacks, tunnels, or high bridges were necessary. They could build anywhere—or more precisely, everywhere. All that was necessary was to make earthen embankments on which the tracks could be raised above the waterline. What an inexpensive way of making railways. While it was clear that flowing water needed to be taken into account, engineers—in their enthusiasm—argued that twenty-five or thirty-year floods were irrelevant to railroad construction.[40] They put culverts into the embankments, but small culverts, useful only for small water flows.

But this was the mighty Ganges. Floods were an ordinary occurrence. The flow of water exceeded engineers' plans. Yet it is the blocking of water that most concerns Iqbal. Railway embankments obstructed and divided river courses, preventing the action of water. The crisscrossing network of railway embankments created compartments in which water became still.

Colonial officials knew that water was a necessary infrastructure for transportation in Bengal, Iqbal explains, and they did try to preserve some of the most important transport waterways. But any waterway that was not classified as key for trade was abandoned; railway embankments cut and stilled the waters. The ecology of the water system changed; the flushing action of the delta was impeded, and waters became more brackish. Farmers could no longer depend on new nutrients and newly emerging land. Water hyacinth entered and took over.

As water hyacinth covered the surfaces of the newly stilled water compartments, travel was stymied except for those who took trains. Fishing disappeared as fish died in the oxygen-poor waters. Diseases involving water-borne organisms spread. Meanwhile, irrigation canals were blocked, and irrigated fields were smothered with water hyacinth. Paddy production dropped rapidly. Bengal changed from a region rich in rice paddies to one dependent mainly on imported rice from Burma. When Japanese occupation cut off the Burma trade, Bengal was in trouble. Famine followed. Amartya Sen has argued that this famine was a human-caused event, a result of unequal entitlements.[41] Iqbal does not deny the significance of unequal access but points to the regional ecological vulnerability in which it operated.

Iqbal follows colonial officials' response to the rise of water hyacinth. Two kinds of responses predominated: First, they commissioned studies to see if water hyacinth might be good for something; second, they passed laws against it. Both were spectacular failures. Both remain major responses to water hyacinth overgrowth around the world.

South Asia continues to be a major site for research on water hyacinth utilization, and it is interesting to read this agronomy literature.[42] Would-be water hyacinth managers have many dreams for this plant: It already behaves like an ideal crop, producing more and more of itself. Why not feed it to cows or pigs or poor people? Why not use it to fertilize fields? Why not ask it to purify sewage? Or at least, why not ask poor people to make baskets out of it? Each of these, and more, have been tried, and you can find the internet littered with promotional sites for each one. But reading the basic agronomy studies is instructive: Water hyacinth does none of these things well enough to bother. Cows and pigs and even people will eat water hyacinth if there is nothing else around. Yet at least the cows and pigs, who are measured, lose weight unless their feed is strongly supplemented by something better.[43] Water hyacinth does have nutrients that could be used on fields, but it's mainly water, so it's difficult to move around, and, once dried, it loses most everything worth contributing. It depresses the growth of some plants, such as tomatoes.[44] The list of failures goes on.

Passing laws against it has been equally unsuccessful as a control strategy. Laws pass the responsibility down the ranks, setting out to punish lower participants who do not work hard to get the water hyacinth out of the way. Iqbal argues that sporadic drives had their moment.[45] Even British officials liked to get out of the office now and then to join public anti–water hyacinth drives. But as soon as the momentum of a particular drive died down, the water hyacinth was back, seemingly unstoppable.

The utter inappropriateness of these responses to the water hyacinth crisis is impressive.[46] In this disconnect, as with Morita's surprise, might an ontological rift emerge between the world-making of colonial officials and the world-making of the plant itself? Water hyacinth has been busy creating its own infrastructure, despite the pleas and plans of both peasants and engineers. Perhaps this is a good time to turn to the plant itself.

Water Hyacinth Agility

Water hyacinth builds an impressive architecture: mat-like structures of plants packed in close together. It is these mats that block the flow of water in canals, kill fish, support mosquito larvae, crowd out paddies, and, in water tanks, pull out the water for their own uses with little left for others. There are two features of the plant's growth that allow these tight mats: first, the ability to send out stolons, producing new plants, and second, the plasticity of growth that allows different plant morphologies. Consider the latter: Water hyacinth plants have quite different forms, from long and skinny to fat and bulbous. A few plants by themselves always produce the bulbous stems that keep them afloat. Squashed together, the long thin morph develops; the plants' sheer mass holds them up. This is a mat-making engineering ability. The plants work together to move from small packs to amassed crowds.[47] Together, they conquer whole bodies of water, making them their own.

What about clonal reproduction, already mentioned in Chapter 7?[48] Evidence supports an argument that, as the predominant form of water hyacinth reproduction, clonal reproduction is an *achievement* of diaspora as much as a precondition. Water hyacinth in diaspora succeed to the extent they produce clones; the plants have moved toward a more purely clonal form. This is seen in the close genetic identity of most water hyacinth outside of their Amazonian homeland.[49] This is not just because only a few plants first made it abroad; there has been enough time to develop diversity from sexual reproduction and local adaptation. Instead, successful water hyacinth in diaspora are those that use their energies to make clones.

To appreciate this requires further immersion into the world of water hyacinth sex. Water hyacinth belong to a botanical family that has an elaborate sexual system to ensure outcrossing—that is, reproduction through the mixing of genetic materials, guaranteeing diversity. With the diversity of sexual reproduction, the plants ensure that their offspring are ready to thrive in varied environments. Here's how they do it: Within this family, each plant has one of three sexual forms. Some have long styles (that is, the female organ that receives the pollen) and shorter anthers (that is, the male part that distributes the pollen). Some have medium-sized styles and one long and one short anther. Some have short styles and mid-sized and long anthers. This arrangement is called *tristyly*. It means that no plant can be pollinated by pollen from its own morphotype. Pollination occurs when a pollen grain from one morph, carried by an insect, encounters the style of another morph. This is ensured by the disparate heights of the male and female parts; in addition, pollen grains from each morph are different sizes, ensuring incompatibility in self-pollination.[50]

Water hyacinth in their South American homelands have just this form of reproduction, which guarantees a variety of forms to colonize new areas. However, plants in diaspora succeed precisely because they have managed to overcome it. In diaspora, the short-styled morph has never been found. The long-styled morph is rare. Only the medium style has been successful. Furthermore, these plants have developed a new trait: They can produce by self-pollination without any insects to carry their pollen. With or without insects, they can reproduce.[51]

Clonal reproduction is one stage further: No sexual recombination need occur. The plants just produce new forms of themselves. How can the plant have abandoned outcrossing, the very mechanism that brought the family forward through evolutionary time? This is a successful diasporic strategy for only one reason: The environments made by imperial engineering are so homogeneous that producing diversity has become irrelevant. Landscape governance projects have produced such uniform environments that the plant needs only propagate itself as quickly as possible to succeed. This interwoven evolution of human engineering and plants is why one can call this biology feral: It is not a domestication, but it is a close engagement between human and nonhuman trajectories. The standardization that modern landscape projects build into the world called forth a plant standardization that could match and outdo engineers' desires.

Clonal reproduction is also the way the plant conducts *infrastruc-*

tural work, that is, how the plant maintains itself in the face of constant human efforts to remove and kill it. Only clonal reproduction can overwhelm the speed and effectiveness of human removal projects. Water hyacinth succeeded as a third infrastructure on the landscapes described earlier only to the extent that the plants' reproductive abilities overwhelmed the efforts of human engineers. Water hyacinth infrastructures depend on the abilities of a few plants neglected in cleanup to reproduce so quickly that mats reform under the engineers' noses—thus too giving the plants time for occasional flowering and seed production, the better to deal with drought and floods.

Changes in the reproductive biology of water hyacinth give a sense of how active and reactive the plant can be in making worlds. The term "ecological engineer" has been used by ecologists not to refer to engineering *of* ecologies (by humans) but rather engineering *by* nonhumans.[52] Beavers build dams, which create ponds for fish. Earthworms tunnel through the ground, spreading nutrients and aerating the soil. Roots hold on to soil and rocks, preventing erosion. These are examples of what organisms do as ecological engineers. Water hyacinth in diaspora, too, is a powerful builder of ecologies. But unlike these other examples, it requires human partners, and not just any humans, but engineers. It builds together with human engineers, even as it thwarts engineers' plans.

How does the activity of water hyacinth make a difference in thinking about history? It is useful to return to the possibilities of doing archival work through landscape infrastructures. It is possible to read *along the archival grain* of each infrastructure to learn the categories it sets out into the world, that is, the way it makes subjects and objects. The infrastructural work of the Panama Canal includes maintaining watershed forest and clearing water hyacinth.[53] The infrastructural work of Panama's "conquest of the jungle" campaign makes farmers, fields, and fallows. Each infrastructure offers us access to an ontological framework, a framework for world-making. The infrastructural work of water hyacinth in the canal zone's Lake Gatun makes mats on the still water. Each infrastructure offers a way of doing history on the landscape. As in Richard Price's separation between Saramaka oral histories, colonial plantation records, and Moravian missionary memoirs, each of the infrastructures of the canal zone tells a separate but interwoven story about the past in the present.[54] Canal administrators show us changing ship sizes; Panamanian farmers show us banana diseases; water hyacinth show us agilities in cloning. Multiple ontologies bulge out of landscapes as archives.

Any given landscape can be understood in relation to the interplay of its multiple infrastructures. This is landscape as throughscape—and landscape history as a series of articulations. Each of the authors this chapter has followed into water hyacinth waters has already found multiple infrastructures, even before considering the work of water hyacinth. Infrastructures, they tell us, are overlaid in palimpsest. There is no reason to allow human exceptionalism to block attention to further multiplicity. This is not only because social relations involve nonhumans as well as humans. It is also because infrastructure selectively advances more than merely human lifeways. Nonhumans may participate in human-planned infrastructures. But they may also participate in infrastructures that work against human plans. Human and nonhuman infrastructural work is entangled—and not in predictable ways. In feral biologies, such as that of diasporic water hyacinth, floating mat infrastructures exist because of imperial engineering even as they get in the way of its purposes. To tell histories of imperial engineering requires nonhuman infrastructures and histories; and vice versa. The fact that they refuse to submit to the same world-making categories becomes a challenge for analysis.

If each of multiple infrastructures offers us a guide to a kind of world-making, each is also a guide to the frames through which pasts and presents might be read in the landscape. Together they make history multiple. It is here that one can see how reading multiply along the grain brings historians back to reading against the grain. Landscapes exaggerate an aspect of all archives: There are always traces of more there than a single world-making project can capture. One can read divergent but through-laid stories on a landscape: Landscapes show us tracks and traces of many historical ontologies. These historical ontologies may be separate, but they are not autonomous; they form within encounters and equivocations. As analysts, we are responsible not only to follow one world-making project but also the cosmopolitics of their interactions. Investigating infrastructural work in palimpsest is one way to do this.

Within landscapes as archives, there are always many infrastructural projects; one storyline cannot hold everything in. This is equally true for human written archives. Reading against the grain can be the attempt to catch the scent of excess in the archives. History opens to the multiple contradictory and constitutive projects—human and not human—that it never fully contains.

Coda: Ways to Play with Water Hyacinth

Perhaps this chapter can hold one more water hyacinth story, offered by anthropologist Kristien Geenen.[55] It is a story of a series of posters in a provincial town in the colonial Congo of the 1950s. The first poster appeared in 1955 when the town prepared for a visit from the Belgian king by constructing modern buildings, announced on posters as "Congo ya Sika" (the New Congo). The second series of posters in 1958 was also marked "Congo ya Sika." But these had a picture of water hyacinth and ended with the admonition: "Let us destroy it."

Someone had decided that the Swahili name of water hyacinth in the Congo would be "Congo ya Sika," since it had never been seen before. This town was not a place whose waters had been invaded by water hyacinth. But colonial authorities, frustrated in their efforts to clear the plant from the waters it had taken over, had moved their campaign to places where there was no water hyacinth, to prevent it from spreading.[56] Local townspeople drew a different conclusion from the posters. This plant must have incredible abilities: After all, it represents the New Congo! The posters asked people to gather any plants they found and hand them to the Belgian colonial authorities; rumors circulated of the huge rewards that would be given to anyone who found the plant. Perhaps the Belgians wanted to steal the plant to bring home the riches it promised before Congolese could benefit. (Even today, rich townspeople are thought to have gotten their start from water hyacinth.) Indeed, the roots looked like the hair of Mami Wata, the water-loving deity who promises modern wealth in exchange for a human sacrifice. The authorities piled up the plants the townspeople turned in, and, when it became clear that residents were trying to steal plants from the pile, they sent soldiers to guard it night and day—thus further promoting the plant's outsized reputation. The bonfire that burned the pile had the same effect: This "New Congo" must be something very, very valuable.

This story reminds readers—delightfully—that not everyone reacts to water hyacinth with the horror of water engineers. People are creative: Water hyacinth can mean different things. Perhaps human exceptionalists will jump up to argue, "This shows that plants are only what we make of them. It's human ways of 'doing' water hyacinth that matter; they *become* through our ontologies." Consider, however, how little contact the people in this story had with water hyacinth. Their ideas were based on a poster with a picture of water hyacinth, not the plant itself. The story tells more about the world-making interactions of people and posters than those of people and plants. This doesn't mean people's relations with plants are limited; but it does remind us to check practices—whether with posters or with plants.

People make worlds informed by distinctive ways of knowing and doing—but these ways are always involved in encounters with the practices of others. The landscapes sedimented in these sometimes-messy interactions are not controlled by one or another world-maker; they arise in the often-unintended coordinations among world-makers. When we consider landscape-making practices, there need be no great divide between human and nonhuman *or* myth and materiality. The key is to watch how they come together in the practices through which varied participants work together. In the nineteenth-century United States, Mormons made wide roads to please God.[57] In Ecuador, Runa hunters communicate with dogs.[58] In the Meratus Mountains in Borneo, celebrations send rice off on a long journey to whisper in the ears of princes and bring back wealth.[59] Each of these practices makes landscape—as it also interacts with the landscape-making practices of roads, dogs, and rice, respectively.

Taking landscape as a record of ontological coordinations offers a different way of doing history than the more common relativisms that grace contemporary anthropology. In most such visions, societies have historiographies—or, like Amerindians according to Viveiros de Castro, "ahistoriographies." Take the position of Marshall Sahlins as an example.[60] According to Sahlins, societies have characteristic ways of doing histories, from kingly transitions to sexual metaphors of reproduction. These ways of doing history are put in jeopardy when in contact with other historiographies, characteristic of other societies. Sahlins's example is the eighteenth-century encounter between the English explorer Captain Cook and the Hawaiians who saw him as the god Lono; this cross-cultural encounter, Sahlins argues, changed Hawaiian ways of making society and history. Sahlins's approach seems useful, although it reifies society; using Gayatri Spivak's term, it is a "strategic essentialism."[61]

But look what happens when we notice nonhumans. Cross-societal encounters are everywhere. Plant sociologists are perfectly happy to say that plants have societies. Human and nonhuman societies encounter each other. But, in this approach, every time you bite into a tomato or pull a weed from your yard, you recapitulate the encounter of Captain Cook and Lono, the Hawaiian god. That's fine but note that it puts cross-societal encounters at the heart of everyone's everyday life—every day, everywhere. Furthermore, the histories of human and nonhuman societies are hardly autonomous from each other. We owe our weeds to ancient domestications; tomatoes came to Europe with the European invasion of the New World. Diasporic water hyacinth histories and colonial engineering histories emerge through their articulation. Once one considers nonhumans and humans, the urge for the autonomy of each historical project withers. Historical ontologies are always cosmopolitan, fully engaged in dialogue with other ontologies. In the search for historical patterns, it is impossible to read along the grain without also reading against the grain—and vice versa. This double reading offers an opening to that "better description" that Marilyn Strathern advised anthropologists to continue to seek.[62]

NINE

Histories of the Future

ANNA TSING

The Anthropocene emerges in irony: Humans are creating a world for the agency of feral nonhumans. Those nonhumans capable of thriving with imperial and industrial infrastructure have taken the reins of history. The world we are building hardly belongs to us anymore; it belongs to them.[1]

How do nonhumans become leading protagonists in emerging histories of the Anthropocene? This chapter takes a closer look at the agilities through which nonhumans come to coordinate with human infrastructures, making histories. Many of the mechanisms discussed in this chapter have already been introduced in *Field Guide*, but the narrative here gains focus and force through homing in on one kingdom: the fungi. Fungi can be some of the most dangerous feral beings on earth, with the ability to wipe out fields and forests, and to extinguish species. They constantly refigure themselves, the better to shape the future. Their force impresses: They allow us to focus on just those agilities that leave us humans in the dust. It is worth thinking through them to revisit the mechanisms that put the Anthropocene in place.

In contrast to Chapter 8, the theoretical work in this chapter proceeds, not through scholarly interlocutors, but rather through the identification of foci for observation. Still, some clarification seems in order. First, scientific observations, like observations in the humanities, are relational—that is, dependent on methods and sites of observation, rather than transcendental truth. This does not make these observations wrong; however, it does mean they can be challenged by other kinds of observations. Second, gathering varied kinds of observations, as this chapter does, offers a practice of "piling" (see Chapter 10) that offers potentially insightful juxtapositions with hegemonic common sense. Marilyn Strathern pioneered this kind of practice for social theorists.[2] Third, the structural analysis of ecological catastrophe offered in this chapter presents an alternative to hegemonic assumptions about the shape of the future, in which the advancement of civilization and technology destroys the natural world by suppressing its agency, bringing it to heel. Instead, I argue that particular kinds of nonhuman adaptabilities, here fungal, shape a world of ecologically dangerous articulations.

Fungi as Figures for More-Than-Human Futures

Some years ago, I wrote a book entitled *The Mushroom at the End of the World: On the Possibility of Life in Capitalist Ruins.*[3] The book told of the expensive gourmet mushroom, matsutake, which grows in daunting landscapes—yet empowers mushroom pickers across the temperate world. Since that time, fungi have come into their own as charismatic objects and as figures with which to appreciate more-than-human livability. Matsutake are mycorrhizal fungi, that is, fungi with symbiotic relations with plants. In these troubling times, mycorrhizal fungi open attention to the importance of interspecies ecologies in the resilience of life on earth. Through their generous metabolisms, which bring nutrients to beings other than themselves, mycorrhizal fungi nurture forests. As figures, they show us the importance of mutualisms in the web of life.

Yet there are also other kinds of fungi, and they, too, are figures for human relations with the environment. Pathogenic fungi—fungi that kill—show us the dangers of the Anthropocene. Working together with human infrastructures, fungi have been responsible for more species extinctions than any other group of organisms.[4] While mycorrhizal fungi have supported the resilience of anthropogenic Holocene ecosystems, such as woodlands, pathogenic fungi have responded to Anthropocene conditions to undermine these same ecosystems, threatening more-than-human livability.[5] Fungi, more than any other kingdom, reveal the structural roots of the biological catastrophe we call Anthropocene.

Fungal Coordinations

Most fungi live together with plants, and coordinations between plants and fungi make the diversity and richness of forest life possible. Plant–fungi coordinations can also involve humans, as Elaine Gan and I argue regarding matsutake growing in human-disturbed forests in Japan.[6] Pathogenic fungi are also involved in coordinations with both plants and animals, including humans. This chapter explores this problem: What kinds of anthropogenic coordinations encourage fungal virulence? How have fungi adapted to the infrastructures of the Anthropocene, becoming some of its most deadly collaborators?

Four characteristics of Anthropocene feral biologies are important to understand the new dangers of fungal pathogens. First, global spread: How do feral organisms move from limited territories to become part of a "contagious" cross-continental Anthropocene? Pathogens, like other organisms, have always moved around. But the speed and scale of intercontinental transfers that have been enabled by European imperial expansion and global capitalism are unprecedented. Long-distance dispersion once was slower and more sporadic, allowing forests, for example, to slowly recover from plagues. Constant and rapid long-distance movement along the infrastructures of industrial trade and imperial conquest have created new forms of destructiveness, and these forms are so novel that we identify this movement as part of the Anthropocene.

Pathogens spread through attaching to Anthropocene vehicles, in the widest sense of the word, thus becoming much more than localized problems. The next part of this chapter offers a particular history with which to consider this problem: the spread of dry rot with imperial ships and wood-framed buildings. Dry rot is not a pathogen, since it only infects dead wood; compared to many Anthropocene stories, this is a benign one. But it vividly illustrates key principles of the Anthropocene proliferation of feral ecologies.

Second, enabling infrastructures: What Anthropocene conditions encourage the emergence of dangerous feral biologies? The existence of pathogens is as old as the living world. There is nothing particularly Anthropocene about the existence of pathogens. And yet, Anthropocene conditions, such as the concentration of genetically uniform organisms in simplified ecologies, encourage both the gathering of pests and pathogens as well as the evolution of new forms. The second part of this chapter explores the problem of feral virulence in relation to industrial monocrops, a breeding ground for fungal pathogens. To give concreteness to this exploration, this section introduces plantation pathogens such as rubber leaf blight, conifer root rot, and southern corn leaf blight.

The third focus is the creation of newly virulent pathogens through hybridization and host switching. Such development is easy when closely related species are crowded together, as, for example, in the commercial nursery trade (for plants) and the global trade in live animals. Emergent species and varieties flourish in the interplay between the simplified ecologies of commercial agriculture, on the one hand, and the crowded multiplicity of the industrial trade in living things, on the other. While some emergent species and varieties are just fine at living with others, this is an opportunity for the development of new forms of pathogenic virulence. The pathogens discussed in the second part will come up again here, with others added such as white pine blister rust and Bd (*Batrachochytrium dendrobatidis*) chytrid fungus.

The fourth part turns to the emergence of fungicide-resistant fungal pathogens. This section mentions several that infect humans, including *Candida auris*, a sometimes-deadly fungus that has taken up residence in hospitals. *C. auris* is practically impossible to eradicate, as it has developed resistance to most fungicides, even in combination with others. Furthermore, different variants of *C. auris* have different drug-resistant abilities, and they can exchange these with each other when they come into contact. How might such a situation have come into being? By saturating land and water with fungicides, Anthropocene agriculture has made a world for fungicide-resistant pathogens. Meanwhile, many ordinary soil fungi—many of which are beneficial to plants, including crops—have been killed off. What is this world we are making?

Anthropocene fungal virulence creates a dangerous world for both humans and many nonhumans. This chapter adds to the articulation theories developed in Chapters 7 and 8 to consider how Anthropocene infrastructures make disease environments.

When Do Pathogens Go Global?

About 1798, there was, at Woolwich, a ship in so bad a state that the deck sunk with a man's weight, and the orange and brown colored fungi were hanging, in the shape of inverted cones, from deck to deck.

—*British Admiralty report*[7]

Between the early eighteenth and the mid-nineteenth centuries, the British navy experienced a serious setback: Its ships began to rot before they could even set out to sea. In his report on this history, John Ramsbottom writes: "The duration of a ship was estimated at twenty-five to thirty years in the seventeenth century, about twelve years from 1760 to 1788, about eight years during the Napoleonic period, dwindling to 'no duration' immediately after Trafalgar."[8] The problem was the dry rot fungus. It was the "death-knell of wooden war ships," solved only by the introduction of ironclad ships such as the British *Thunderer* in 1863.[9]

The following is clear: Humans did not invent destructive fungi. Destructive fungi have existed a lot longer than humans. But Anthropocene infrastructures have succeeded in the round-the-world spreading of some fungi that otherwise would have remained in circumscribed places. It is that spread, in many cases, that has caused all the trouble. It is that spread that counts, in this section, as a "feral biology."

Dry rot (Figure 35), caused by the fungus *Serpula lacrymans*, is not a pathogen; it eats dead wood. The fungus is only important if it happens to infect a valued piece of dead wood, such as a ship or a house. It makes a good protagonist for this part of the story only because it shows how fungi have been transported around the world—in this case, as part of imperial ships. In its natural habitat, *S. lacrymans* is a specialist that competes poorly with other fungi; it lives on dead wood in shade at the cool, dry tree line.[10] In the protected environments of human construction, however, it thrives—and proliferates.

The distribution of *Serpula lacrymans* has confused geneticists. Because one variant is so widely distributed—in human habitations—across Europe and North America, the search was on for its wild relatives. Several scattered relatives were found, but the closer researchers looked, the more different these relatives appeared from the kind that lives in ships and houses, which they dubbed the "cosmopolitan" vari-

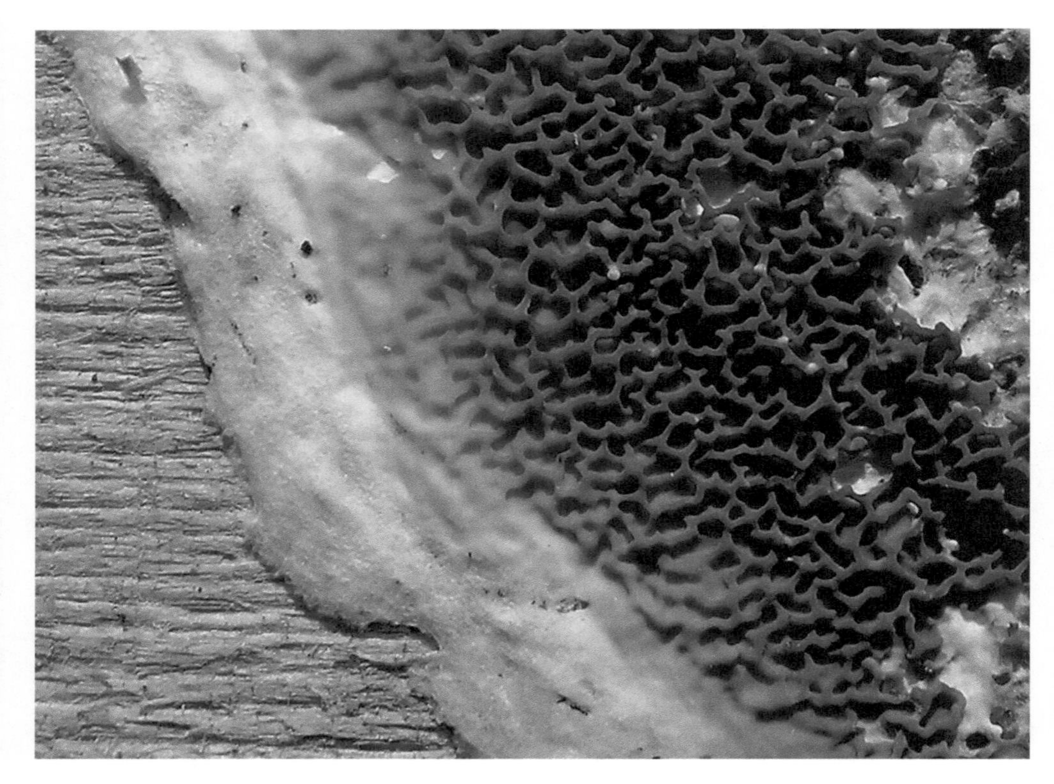

FIGURE 35. **Close-up of dry rot fruiting body.**
EPP

ant.[11] There are indeed wild dry rots in mainland Asia, but they are rare, parochial, and specialized. Furthermore, the cosmopolitan variant has a very small genetic range, suggesting recent introduction. Geneticists are wary of more-than-human histories, which cannot be tested with their methods. However, most authorities have come to accept the idea offered by South Asian "fungus hunters" in the early twentieth century: Dry rot was transported around the world by the British use of logs from the western Himalayas.[12] They quote Ramsbottom:

> Foreign timber was often floated down rivers and then immediately loaded into the confined holds of timber ships, an ideal arrangement for fungal infection; the logs were sometimes covered with fruit-bodies before they reached the dockyards.[13]

From British ships, dry rot toured the world, establishing itself particularly in areas of British settlement, such as North America and Australia.

Dry rot thrived with the vernacular architecture of British empire, which involved wood along with nonorganic materials. Dry rot has a method to move across non-wood materials, such as plaster. It cleverly makes long cord-like rhizomorphs (root-like fibers made from fungal tissue) to explore potential food sources.[14] It excavates useful minerals from plaster. It travels from one part of a house to another part, and from house to house. Wood-framed cellars are perfectly designed for *Serpula*, and wherever such cellars became popular and the temperatures were cool, dry rot came along. Dry rot was a companion of British colonial culture. In learning to occupy a world where the sun never set, it changed from an aid to local forest succession—making nutrients available for a new generation of plants—to an agile cosmopolitan and a dreaded house guest.

Dry rot responded to Anthropocene proto-industrialization: the global deployment of standardized products. Its history shows how feral biologies emerge with new forms of travel. Ships took dry rot from continent to continent; houses provided delightful new habitats for the fungus. But transportation is not the only form of Anthropocene attunement through which feral biologies thrive. To explore new capacities for growth and reproduction, it is useful to turn to the production of resources in simplified, disciplined ecologies, that is, industrial monocrops. Such simplifications have been introduced earlier in this book. But fungi urge another look: Fungi are that kingdom most able to zero in on that simplification not only to destroy whole crops at once, but also to spread beyond the commercial farm.[15]

Industrial Monocrops Are Feral Incubators

Annosum root rot, a white rot resulting from a group of cryptic species of the genus *Heterobasidion*, is the most feared resident of modern conifer plantations. From the perspective of commercial owners, it causes billions of dollars of damage every year. Unlike *Serpula*, *Heterobasidion* can feed on live wood as well as dead, and it is virulent enough to take down whole plantations. But *Heterobasidion* does different things on live and dead wood; its ability to switch back and forth is key to its ability on the plantation to transform from humble fungus into serial killer (Figure 36).[16]

Here is the story: *Heterobasidion* spreads in two ways. In primary infection, a spore germinates on a freshly exposed stump; it wants dead wood. In secondary infection, mycelial growth moves from the live roots of one tree to a closely related other through the natural grafts formed by roots under the soil. Outside of industrial monocrops, neither infection route causes much forest damage, and *Heterobasidion* has trouble moving very far. Instead, it stimulates succession by killing off a few trees, allowing others to replace them.

In an industrial monoculture, however, *Heterobasidion* has an enhanced ride. Stumps are located near live trees, providing routes for initial infection. The live trees, arranged in closely spaced rows, are

FIGURE 36. **Fruiting body of *Heterobasidion* root rot on a tree stump.**
GERHARD ELSNER

FIGURE 37. ***Heterobasidion* spores germinate on stumps and then spread through natural root grafts.**
FEIFEI ZHOU

genetically close and thus likely connected by a mass of natural root grafts—routes for further infection (Figure 37). The industrial monocrop makes a new world for the fungus: a world where nothing stops it. The variety and patchiness of woodlands, which kept *Heterobasidion* humble and small, is replaced by one great womb for *Heterobasidion* spread and reproduction. From stump to roots, and from one root to another, the conifer plantation seems designed for the unlimited reproduction of the fungus. Indeed, *Heterobasidion* is not alone: Think of any industrial monocrop, and there is a fungus that has remade itself as its champion killer.

A devastating incidence of southern corn leaf blight (SCLB) makes this point even more directly: Fungal virulence is stimulated by the political ecology of industrial monoculture. The narrow genetic stock that makes industrial monocrops easy to work with for capitalist economies of scale also makes them breeding grounds for fungi able to develop variants to infect just that genetic stock. Consider corn, where breeding practices created plants so genetically similar that their height and time of ripening is standardized, facilitating mechanical harvesting. The genetic similarity of the crops has meant that fungi that evolve the ability to infect one plant can, indeed, infect them all.

Cochliobolus heterostrophus, the fungus that causes SCLB, was a minor pest until 1970, when it managed to kill 15 percent of the corn crop of the United States, as well as corn across all those parts of the world supplied by U.S. hybrid seeds (Figure 38). Journalist Jack Doyle described the episode with the alarming drama the blight deserves:

> The new fungus moved like wildfire through one corn field after another. In some cases it could wipe out an entire stand of corn in ten days. Moisture was a key factor; a thin film on leaves, stalks, or husks was all the organism needed to gain entry to the plant. Within twenty-four hours it would start making tan, spindle-shaped lesions about an inch long on plant leaves, and in advanced form would attack the stalk, ear shank, husk, kernels, and cob. In extreme infections, whole ears of corn would fall to the ground and crumble at the touch.
>
> The fungus moved swiftly through Georgia, Alabama, and Kentucky, and by June its airborne spores were headed straight for the nation's Corn Belt, where 85 percent of all American corn is grown. By this time, however, unsuspecting Corn Belt farmers had already

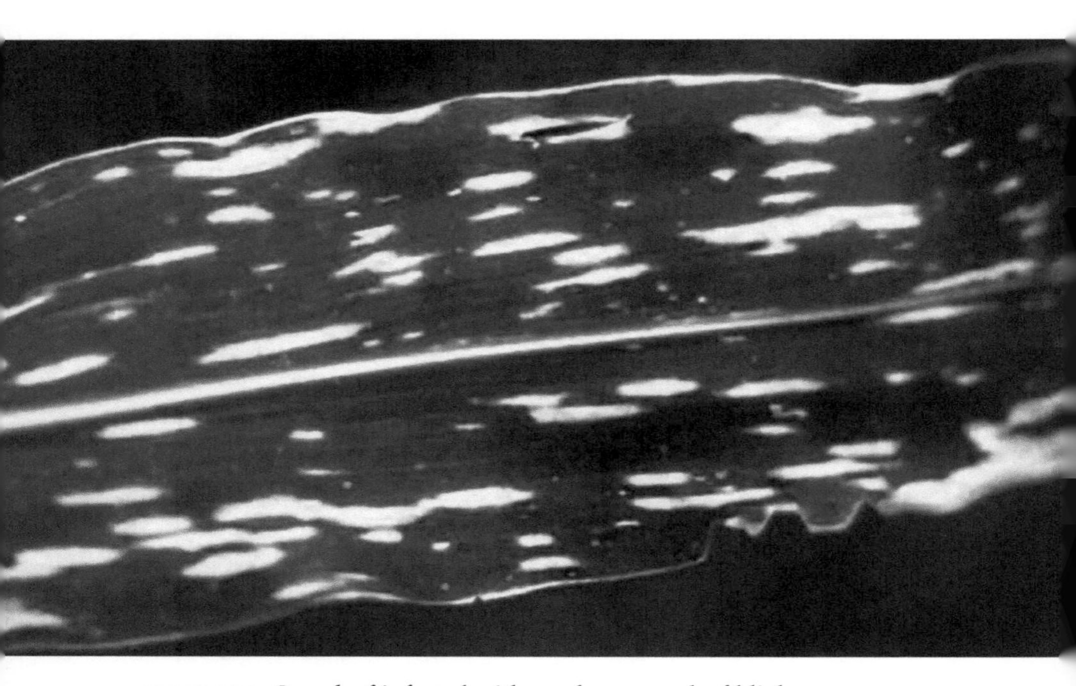

FIGURE 38. **Corn leaf infected with southern corn leaf blight.**
DAVID B. LANGSTON, UNIVERSITY OF GEORGIA, BUGWOOD.ORG

planted their crops and were largely unaware of the bitter harvest headed their way.[17]

The astonishing rise of SCLB turns out to have been enabled by a genetic modification that caused male sterility in corn plants. Corn is pollinated by the wind; this means that farmers cannot control the parentage of their corn ears as long as pollen from varied corn plants is around. After the development of high-yielding hybrid races of corn in the 1930s, farmers who planted hybrid seed hired workers to remove the male flowers of corn, called tassels, by hand. At midcentury, however, U.S. agronomists had found an abnormal corn plant that was already sterile for male flowers, and they developed it through back crossing into their hybrid seeds, which became widely available in the late 1960s.[18] Since industrial planting practices homogenized seed sources, most all the corn in the U.S. Corn Belt, and much corn in other countries, used this single genetic modification. This homogeneity laid out a feast for the right fungus. When *Cochliobolus* developed a "T variant" readily able to infect plants with this modification, every plant was stricken. These were the conditions for making a plague.

Agronomists quickly understood that the fungus was keyed particularly to this single genetic line. Unfortunately, seeds for the 1971 crop had already been sold both in the United States and around the world; thus another year of blight ensued. Subsequently, the male-sterile breeding line was withdrawn, and T-variant *Cochliobolus* was no longer a problem. Yet this quick action was only a success in sustaining the destructive business practices of capitalist agriculture. While many scientists spoke of the need for maintaining genetic diversity in crop fields, this did not happen. Another strain of genetically homogeneous corn became dominant, setting the stage for another fungal blight in the future, as soon as a fungus finds the way to infect this one. This has been the practice, too, for the many fungal rusts and blights infecting other commercial crops: Offer lip service to the importance of genetic diversity; find a resistant variety; and then continue monocropping practices as before. If 1970 and 1971 seem ancient times, consider the still-spreading yellow-stripe wheat rust: "Losses of 40 percent can be common with some fields totally destroyed."[19] Agronomists have responded in a race to develop new wheat varieties, hoping they will stay ahead of rusts; but fungi catch up. The point of critical discussion of the Anthropocene has been to block such business-as-usual frameworks. To take seriously the planetwide dangers of the Anthropocene might mean, instead, to question industrial monoculture itself.

What Does It Take to Become an Anthropocene Killer?

Industrial monoculture is useful to think about the scale of fungal plagues. Some fungal pathogens are limited to particular crops, while others spread widely and jump from host to host. The difference matters: While the former is merely an annoyance to plantation owners, the latter destroys living arrangements for many ecological communities. The difference often depends on commercial travel trajectories, as these are involved in the spread of industrial plantations, with their pathogens.

An example of a pathogen that has not (yet) traveled is rubber leaf blight, *Microcyclus ulei*, which continues to be found only in the Amazonian homeland of the rubber tree.[20] When Henry Ford opened his failed rubber plantations in Brazil, which he called Fordlandia, in the 1920s, he stimulated the rapid reproduction of rubber leaf blight, which destroyed his rubber-production dreams.[21] *Microcyclus ulei* did not have to develop a new variety; merely planting genetically homogeneous rubber trees so close together that their leaves touched allowed conidia (asexual spores) of the fungus to spread from tree to tree, killing off the plantation. The presence of the fungus in Brazil accounts for the fact that rubber plantations continue to be found only in Africa and Southeast Asia, where the fungus has not yet traveled.

In contrast, the root rot mentioned above, *Heterobasidion*, has begun to travel through the movements of colonialism and industry. As conifer plantations are carried to new parts of the world, so is *Heterobasidion*. Under conditions of global resource mixing, plantation conditions create new powers for *Heterobasidion*: They allow variants to get together, forming new fungi with new powers. *Heterobasidion* has an extremely long genome with many transposable elements, and the new hybrids effected by resource mixing seem to have activated such elements in ways that allow them to pick on new hosts. In western North America, for example, the bringing together of *Heterobasidion irregulare* and *H. oc-*

cidentale in pine plantations has created the ability to attack sequoia, which was previously unaffected. The mixing of different variants of *Heterobasidion* has allowed the fungus to spread across non-plantation landscapes, killing new kinds of trees. In Italy, the appearance of North American *H. irregulare* among native European *H. annosum* has managed to bring the pathogen into oaks—not a conifer at all. As some researchers suggest, this new infectivity "may allow the pathogen to reach pine stands in the highly fragmented landscape of central Italy."[22] *Heterobasidion* has begun to map out a truly global terrain—by forming combinations that attack new hosts and reshape landscapes both inside and outside the tree plantation.

The ability of *Heterobasidion* to infect new hosts, outside the tree plantation, stands in for a major fungal pathogen issue: hybridization and host switching as modes of emerging virulence.[23] This is a problem for the Anthropocene; commercial nurseries and the global trade in living organisms stimulate hybridization and host switching. As Eva Stukenbrock explains,

> Hybridization occurs more readily between species that have previously not coexisted, so-called allopatric species. Reproductive barriers between allopatric species are likely to be more permissive allowing interspecific mating to occur. The bringing together of allopatric species of plant pathogens by global agricultural trade consequently increases the potential for hybridization between pathogen species.[24]

Jasper Depotter and his colleagues explicate the relationship between hybridization and host switching in Figure 39.[25] The blue chromosome on top represents a particular fungal lineage, which infects the green plant. When it crosses with a related species, whose chromosome is shown in red, some of the offspring are just like the blue parent, while some (shown by combinations of blue and red) may be more aggressive as pathogens. Some offspring may develop the ability to infect alternative hosts (red plant at bottom right).

Commercial nurseries, as well as plantations linked across continents, are ideal sites for pathogen emergence from these dangerous hybridizations. When plantations are combined with global resource transfers, the feral organisms that rise up from these arrangements may

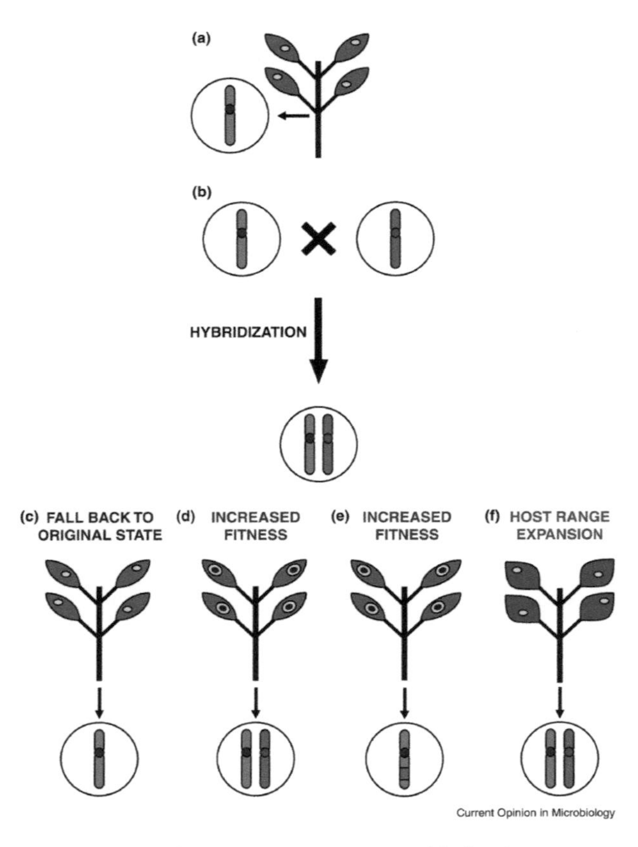

FIGURE 39. **Evolutionary trajectories of hybrid pathogen genomes.**
JASPER DEPOTTER ET AL.

gain superpowers that make it almost impossible to imagine blocks and controls.

Even without hybridization, the plagues of industrial monocrops can spread beyond the plantation—for example, when landscape managers confuse the plantation and the world. It is worth spending some time with what happens when woodlands are treated as if they were a tree plantation. One historical example involves the spread of white pine blister rust, *Cronartium ribicola*, in the United States, where anthropogenic woodlands are understood as forests (Figure 40).[26]

White pines are indigenous to North America, and in many places they grow quickly on abandoned fields. The U.S. timber industry began

FIGURE 40. **Fruiting bodies of white pine blister rust.**
MAREK ARGENT

with eastern white pine, and when those white pines ran out, the industry cut the Midwest, and then, when those white pines ran out, other pines in the northwestern Inland Empire. At that point—in the late nineteenth century—white pine plantations on cutover land emerged as an economic crop. From this point on, plantation pines and naturally occurring white pines intermingled in pine country across the United States. The confusion between tree plantation and forest became a mainstay of the U.S. timber industry. Forests were managed as tree plantations, and cutover lands became both tree plantations and forests. Blister rust was introduced, managed, and spread through this confusion.

Plantation growers wanted seedlings—but American nurseries were getting better prices for ornamental plants. Planters looked to Europe, where American white pine had been introduced. Tariffs were removed, and, by 1909, millions of white pine seedlings were pouring into the United States from low-cost nurseries in central Europe. White pine is not indigenous to Europe, and the nurseries mingled introduced pines from many places. Thus, these cosmopolitan nurseries contaminated their seedlings with an Asian fungus: white pine blister rust. As soon as they saw it killing their pines, Americans were alarmed, and by 1912,

Congress passed its first quarantine law to stop this trade. But they were closing the barn door after the horse had bolted.

Planting imported seedlings in the midst of naturally growing trees is a good way to spread plant diseases. This is treating the forest as a tree plantation, with its feral effects. At the same time, the forest resists the management strategies that might be used in a real plantation, where the vegetation is simplified to meet the needs of human managers. This conundrum—that plantation practices in forests spread diseases while also resisting plantation management practices to control disease—suffuses U.S. Forest Service attempts to stop white pine blister rust.

How could this fungus be controlled? Forest Service managers thought they had a chance to stop the disease because of white pine blister rust's unusual life cycle, which requires two hosts, not just white pines but also gooseberry and currant shrubs of the genus *Ribes*. By destroying all *Ribes* shrubs, managers thought, the fungus could be deprived of a key part of its reproductive cycle. This is what U.S. control efforts set out to do: Eradicate *Ribes*. At first, they thought this was easy work, involving garden-grown black currants. But *Ribes* grows in many forms across the United States. In the West, it is particularly diverse and plentiful in natural forests. Eradication workers could not limit themselves to gardens, or even plantations; they had to comb every forest, removing every shrub.

Before the spread of industrial toxins, plantation problems of all sorts, including diseases, were solved with massive applications of human labor. This made sense within the history of plantations, originally developed as a feature of colonial conquest, which made labor cheap. But to apply this formula to forests asks for trouble. Forests are complex and heterogeneous; they are not easy to control. Labor can be hard to mobilize and uncertain in its effectiveness. Local residents can get in the way. The campaign against white pine blister rust illustrates all these problems.

In different regions, the campaign required different kinds of work and workers. For example, in the southern Appalachians during Prohibition, moonshiners were hired to direct the work crews; otherwise, the "shiners" would effectively block the work.[27] In contrast, in the U.S, Northwest, the problem was not local residents, but rather the forest itself. Here the forests were extensive, the terrain difficult, and the *Ribes* plentiful and diverse. Two developments made eradication imaginable. The first was the bulldozer, which was used in streambeds to uproot everything—not just *Ribes* but all plants. The destroyed areas were planted to grass, with the dream, as one officer explained, "to convert a wasteland of brush into a revenue-producing hay field."[28] The second development was the Depression. Once huge crews arrived from the Civilian Conservation Corps and the Works Progress Administration, real advance seemed possible. And yet this development is just what shows the local specificity of the whole plan. Once World War II loomed, the labor force disappeared. The eradicators tried to make do with prisoners and Mexican migrants, but this was the beginning of the end.

Furthermore, eradicators belatedly learned that making the world into a plantation was part of the problem. Logged-over areas turned out to be the very best places for *Ribes* to flourish; thus, every time their crews managed to save a healthy stand of pines, the timbering of the pines encouraged more *Ribes*, and thus more blister rust. Fighting the fungus by making the forest more plantation-like merely encouraged the fungus. As in Fordlandia, eventually everyone gave up. And U.S. forests, especially in the Northwest, have less white pine. Whitebark pine has almost been wiped out.[29]

The U.S. Forest Service made an all-out effort to control blister rust. Yet not only did they fail to stop it; they also spread it. In the process, they highlighted the contradiction: Disease managers aimed to use plantation techniques, but far from eradicating the problem, they exacerbated it.

White pine bister rust spread through the industrial nursery trade. Such plagues are not just a feature of the past; the trade encourages the development of new diseases as well as spreading old ones. Consider

the phytophthoras. Phytophthoras are not fungi, but they borrow their abilities to get inside plants from the fungal kingdom.[30] New species of phytophthoras, infecting new hosts, are constantly emerging in commercial nurseries. The new pathogen *Phytophthora alni*, for example, kills European alders.[31] The nursery trade then spreads these diseases around the world. Forest pathologist Matteo Garbelotto describes how *Phytophthora ramorum* was introduced into California in the 1990s, where it caused the epidemic called sudden oak death.[32] California woodlands are losing whole populations of trees—and spreading diseases elsewhere through continuing shipments of plants and soils.[33]

Lots of life forms cross borders these days. It's easy to imagine that such transfers are just a natural part of human travel and trade. Perhaps, we might think, they are a necessary concomitant to freedom. After all, are we all supposed to stay home? This response, however, neglects the conditions under which feral biologies develop and spread. This is not just travel: It is industrial travel. If this is freedom, it is the freedom of capitalists to change and destroy life.

Capitalist freedom also leads us to the mass death of frogs and other amphibians. The frog part of the plague, described by Jonathan Kolby and Lee Berger, is caused by a fungus called *Batrachochytrium dendrobatidis*, popularly known as Bd.[34] A global panzootic lineage has circulated in recent years through the commercial movements of frogs.[35] For a number of years, researchers debated where the pathogen originated;[36] the most recent research of which I am aware suggests Korea.[37] However, the source may not be as important as the method of its spread: commercial trade. "An expansion of BdGPL [the Global Panzootic Lineage, which is the really virulent and widespread form] in the 20th century coincides with the global expansion in amphibians traded for exotic pet, medical, and food purposes."[38] Commercial shipping crowds frogs together under less than healthy conditions. Diseased frogs are released in new places, and the frogs carry the fungus to new populations of frogs at their destinations.

Consider this assessment by Kellie Whittaker and Vance Vredenburg:

> *Bd* may be responsible for the greatest disease-caused loss of biodiversity in recorded history. Over just the past 30 years, *Bd* has caused the catastrophic decline or extinction (in many cases within a single year) of at least 200 species of frogs, even in pristine, remote habitats. These rapid, unexplained declines have occurred around the world. While diseases have previously been associated with population declines and extinctions, chytridiomycosis is the first emerging disease shown to cause the decline or extinction of hundreds of species not otherwise threatened. Currently over 350 amphibian species are known to have been infected by *Bd*.[39]

Later in their review, they discuss beyond-amphibian consequences:

> Researchers in the NSF-funded TADS project (Tropical Amphibian Declines in Streams) have been investigating what happens to stream communities when frogs and tadpoles are no longer present. Once the frogs and tadpoles die off, algae grow, and nitrogen levels change, with cascading effects both up and down the stream food web. Frog-eating snakes have gone extinct while other snakes have increased. In rural West Africa, Mohneke and Rödel point out that alteration of freshwater ecosystems resulting in the loss of larval anurans may have significant consequences for both humans and cattle; the loss of tadpoles impacts many aspects of freshwater ecology, with consequences potentially including an increase in malaria and a significant decrease in stream health.[40]

Two comparatively immune commercial species have been implicated as carriers: African clawed frogs and American bullfrogs.[41] Both are mules, spreading the epidemic to others. The American bullfrog has been shipped around the world for rural development enterprise. As mentioned in Chapter 3, the promotion of American bullfrog farming in many areas of the Global South has allowed bullfrogs to escape from their farms to infect indigenous frogs across the countryside. As Nathan Snow and Gary Witmer put it, American bullfrogs have a biology "made for invasion."[42] They travel everywhere; they eat everything; they reproduce avidly; and they are bigger than most other frogs and easily expand into their spaces. American bullfrogs not only eat other frogs but also infect the survivors with Bd.

The most frightening thing about Bd is that it kills frogs in remote areas, far from the frog farm. Bd makes use of the flow of water and the sociality of frogs to find its way to places we can hardly think to anticipate, much less control. Furthermore, frogs travel the world not only as commercial cargo but also as hitchhikers in other industrial supply chains. Asian toads traveled to Madagascar hidden in mining equipment.[43] Other aspects of industrial supply chains make frogs more vulnerable. Global climate change makes the world hospitable for Bd. Pesticide exposure makes frogs easier to kill.[44]

At the intersection of multiple capitalist chains, Bd has become a global menace. The resulting plague is a major extinction event of our times. Our grandchildren will inherit a world deprived of amphibians.

Saturating the World with Fungicides Makes Way for More Powerful Killers

> TR4 [causing *Fusarium* wilt in banana plantations] is a soil-borne fungus that is well adapted to long-term survival in soil. . . . The pathogen can be moved in infested soil by both humans (farm machinery, vehicles, shoes and clothing of farm staff) and animals (movement greater on sticky clay soils). . . . [T]he use of herbicides is likely to encourage rapid colonization of senescing plant material, thus increasing inoculum potential. . . . TR4 is virtually impossible to eradicate.[45]

The simplified ecologies of commercial agriculture and livestock rearing gather fungal pathogens and create opportunities for the rapid evolution of new ones. Saturating the environment with fungicides exacerbates this situation, encouraging the development of resistance—and thus new forms of virulence.[46]

Fungi are flexible and opportunistic. Many of the deadliest are capable of rapid evolution, whether through mutations, hybridization, gene exchange, or relations with new hosts. They take advantage of new resources and new niches to remake themselves to survive and thrive. Meanwhile, the range of poisons we can create to knock back fungi is limited, because their metabolisms are similar to ours. Most current fungicides take advantage of cell-wall chemistry differences between fungi and animals. But this offers a more limited set of options than those available to fight bacteria. Fungicide resistance cannot easily be overcome by the development of new fungicides. Even the cocktails of fungicides described by Alyssa Paredes and discussed in Chapter 1 have a hard time keeping fungi off the plantation.[47]

As mentioned in the quotation at the beginning of this section, banana plantations are in trouble, attacked by *Fusarium* fungi. *Fusarium* has mechanisms to survive fungicides. The variety of *F. oxysporum* called TR4 creates asexual spores that persist in the soil, germinating to cause further infection. Plantation bananas all over the world have died in droves. "There are currently no commercial banana cultivars resistant to TR4."[48]

Many *Fusarium* have an intrinsic resistance to fungicides. They also have picked up new forms of resistance after exposure to fungicides.[49] We can't depend too heavily on differences across species and subspecies in this genus because—most excitingly (to a biologist) and most dangerously (to any organism about to be infected)—they have shown themselves capable of horizontal gene transfer, that is, organism-to-organism transfer of genetic material, without sexual reproduction. Entire chromosomes have been transferred across varied strains of *F. oxysporum*, indeed enough genetic material to account for one quarter of the genome.[50] Non-pathogenic strains become pathogens through such transfer. It seems likely that both intrinsic and secondary resistance can also be transferred.

Don't be too dismissive about the plight of the bananas. Some strains of *F. oxysporum* infect humans. Indeed, approximately ten species complexes of *Fusarium* have been found infecting humans: "Among these complexes, members of the *F. solani* complex are the most common and virulent (comprising approximately 40–60% of infections), followed by *F. oxysporum* (~20%), *F. fujikuroi* and *F. moniliforme* (~10%)."[51]

Fusarium species are heralded as geniuses in trans-kingdom pathogenicity. (Humans are in the animal kingdom; bananas are in the plant kingdom.) Pathogenic strains switch hosts, developing further agilities. Resistance to fungicides is only one of many abilities of this fungus.

Other soil fungi have related stories. *Aspergillus* are a group of soil molds that eat dead plant material; they are not, ordinarily, pathogens. They mostly reproduce through clones, but they are capable of sexual reproduction. At least some have characteristics of "panmixia"; in other words, they can mate across all boundaries, quickly spreading new traits—such as virulence and fungicide resistance—across regions and environmental niches.

Aspergillus came to medical attention when an increasing number of infections to humans were reported—and increasingly, too, were found untreatable through medical fungicides. Like *Fusarium*, they had learned to jump kingdoms to infect living humans as pathogens. One set of strains of *A. fumigatus* appears to have gained resistance to fungicides in response to medical treatments; these strains are known among patients who had been prescribed long-term doses of fungicides. This is one way to develop feral agilities, here resistance. But there is another route; another set of strains of *A. fumigatus*, with a different genetics of resistance, appears among patients with no previous experience of antifungal treatment. This set of strains developed in response to antifungal poisons in the environment.[52]

This latter set of strains is the group of pathogenic fungi for which connections to the rise of the use of fungicides in commercial agriculture are most clear. The meteoric rise in fungicide use in agriculture corresponds closely in place and time to the rise in fungicide-resistant *A. fumigatus* infections; the same resistant strains found in human patients were found in flower beds.[53] The infective *Aspergillus* were resistant to the very azole fungicides most used in commercial nurseries and fields, and in paint and coatings, wallpaper paste, clothing, and wood preservation.[54] These azole fungicides pervade not just agricultural soil but wastewater and sewage sludge. *Aspergillus*, saturated in azole fungicides, developed pan-azole resistance by 2007.[55] A soil fungus became a lung fungus—and one that is almost untreatable.

The infective yeast *Candida auris*, discussed by Alex Liebman and Robert Wallace, hit the medical world by storm because no one knew where it came from.[56] *C. auris*, first identified in 2009, is considered an emergent pathogen—and it is resistant to most fungicides. One of the strangest things about *C. auris* is its almost simultaneous emergence through different routes in four different places—and with four different genetic signatures. What happened around the world to allow this dangerous fungus to emerge?

C. auris distinguishes itself by its ubiquity in hospital settings, where no one can get rid of it.

> *C. auris* contamination has not only been seen on bed rails, bed pans, mattresses, linen, pillows, furniture, door handles, flooring, walls, radiators and windowsills, it can even spread as far as bathing areas, sinks, mop buckets and cleaning equipment. . . . Medical equipment that comes in contact with the patient also gets readily contaminated, for example, temperature probes, blood pressure cuffs, glucometers, housekeeping carts, alcohol gel dispensers, dialysis equipment, ultrasound machines, computer monitors, keypads and cell phones.[57]

This ubiquity suggests that saturation with antimicrobials in the hospital—as both medicines and as cleaning products—may be one source of *C. auris*'s resistance.

This hospital story forms just one part of what Arunaloke Chakrabarti and Prashant Sood call "the crucial ecotoxicological disruption . . . the overwhelming saturation of our biosphere with antibiotics and antifungals."[58] Liebman and Wallace agree.[59] They argue that *C. auris* is a "factory fungus," that is, a concomitant of the ecological simplifications and toxic chemicals of commercial agriculture, like the pathogenic forms of *Aspergillus* and *Fusarium*. Might we be actively building a world for fungal pathogens with resistance to fungicides?[60]

The More-Than-Human Anthropocene

Let me return to the effect of fungal pathogens on woodlands, as it shows patch-to-planet dynamics. Losing woodlands is important because of the long historical association of people and woodlands. Most traditional farmers depended on woodlands, whether in agroforestry rotations, as in shifting cultivation, or as woodland-field complexes, as in most settled-field small farming. Woodlands once provided daily necessities for farmers, including firewood and charcoal; they also provided nutrients for fields, whether indirectly through livestock grazing, turned into manure, or directly through green manure.[61] In Europe, the familiar trees of city and countryside—such as pines, oaks, beeches, limes, ash, and alders—are the legacy of anthropogenic landscape making for many generations.[62] Woodlands continue to be both resources for and accompaniments of human livelihoods; this is not about rare species. Humans have always lived with woodlands. Humans need them. Suddenly, the common trees of woodlands are disappearing—hit by the pathogens of the industrial nursery trade.[63]

One important effect is in the emerging carbon economy, in which capitalists have proposed to balance CO_2-producing fossil fuel–burning with CO_2-depleting tree planting. This plan sounds fine–until you realize that tree planting, unless done very carefully, can be perilous, endangering whole forests and entire species of trees. Tree seedlings grown in low-cost industrial nurseries carry diseases, and planting these seedlings around natural forests spreads the diseases far beyond the planting site. The death of trees due to commercially imported diseases easily exceeds the sum total of planting. The capitalist carbon economy will fail in its ostensible goal of offsetting carbon dioxide production as long as it depends on the false basis of *adding* to the planet's tree cover through planting seedlings from industrial nurseries.

Algorithms of tree cover in the Anthropocene generally look for the effects of climate change on where trees can grow. Rarely do they include the effects of commercially spread tree diseases. As long as this is the case, policy and planning will be just plain wrong. Capitalist solutions will be part of the problem.

Through multispecies programs such as these, pathogenic fungi have become collaborators in building a planetary Anthropocene without easy solutions. Perhaps this is obvious; they trouble our health, our crops, and our favorite companions of other species. How best to track these dangerous effects, which reach from patch to planet? Read with the previous two chapters, this chapter forms part of a trilogy, developing theories of articulation for histories of the Anthropocene. Together the three chapters show how human and nonhuman historical trajectories are brought into friction—that is, the rubbing together of varied trajectories that makes worlds.[64] These chapters offer tools for crossing lines between the humanities and sciences in telling more-than-human histories.

Another theoretical legacy that is helpful for understanding this material is Karen Barad's theory of "intra-action." Using Barad's terms, infrastructures are the experimental apparatus necessary for the intra-action she postulates as making materiality within encounters. Intra-action moves analytically beyond interaction, Barad argues, in allowing the nature of material beings to find shape in the encounter; rather than two pre-made entities merely touching, something is created. In contrast to theories of communication and semiosis as the basis for understanding human–nonhuman touching, intra-action allows attention to the *material* responsiveness of encounter. More-than-human histories involving feral biologies are often scenes of infrastructure-mediated intra-action.[65]

Intra-action seems fair game as a sister theory to articulation. Both focus attention on material emergences in the friction of encounter. Both refuse the analytic separation of form and material. Both dwell in what Barad calls "onto-epistemologies" rather than communication scenarios to understand more-than-human becomings. This book shows how feral biologies emerge from encounters with human infrastructure as an "experimental apparatus" that shapes materiality. *Candida auris*, for example, is an emergent species, born from encounters with the Anthropocene apparatus. Attention to the intra-actions in which material forms allows an appreciation of the historicity of Anthropocene biologies.

Pathogenic fungi are finding a place in the patches and corridors of the Anthropocene, making them ever more dangerous. The simplifications and chemical toxins of industrial farming gather pathogens and allow new ones to emerge. The corridors of industrial shipping carry dangerous fungi across continents, spreading them and allowing new and dangerous experiments in host switching. Crowded nurseries and livestock enclosures gather materials from around the world, aiding hybridization, and producing new, more virulent pathogens. If we sit back, they will be the future.

FIGURE 41. ***Yätj Garkman*** **(Evil Frog).**

RUSSELL NGADIYALI ASHLEY

PART IV
Epistemics

What forms of knowledge production are most useful to describe a patchy Anthropocene? In this section, we make the case for a "patchy" epistemics, that is, for ways of assembling knowledge that make use of a diversity of ways of observing, recording, and participating in Anthropocene worlds.

This part of the book explores three ways of building knowledge through diversity. As we will demonstrate, this approach rests on a set of core commitments: (1) think in, and from, place; (2) embrace a variety of descriptive modes; and (3) cultivate connections across difference. In Chapter 10, we show how our understanding of the Anthropocene can be enriched by thinking with and across patches. For this exposition, we show how *Feral Atlas* field reports, generated through multiple kinds of empiricism, enabled the curatorial team to develop an appreciation for "piling" as a knowledge-building practice.

In Chapter 11, Feifei Zhou asks how architects and designers might incorporate a patchy epistemics in their projects. In *Feral Atlas*, multiple modes of world-building were represented in the collages that explain Anthropocene Detonators. Meanwhile, in the world, she argues, building projects sensitive to vernacular and more-than-human modes of being must consider "unbuilding" as part of their practice. Chapter 12 continues our journey "beyond piling." Jennifer Deger writes of her on Country collaborations with Yolŋu Aboriginal people of northern Australia. Rather than pushing Yolŋu analysis and commentary into an established academic mode, Deger and her Yolŋu colleagues have developed new forms of knowledge and aesthetics.[1] This is, in their terms, a "yuṯa anthropology," a practice intended to bring once distant and discrete worlds into relationship in ways that are meaningful and generative for all involved.

Our editorial team has not solved all epistemic problems. For example, we haven't addressed the thorny problem of how to handle the powerful modern spirit beliefs of Europeans and the European diaspora. Anti-science cosmologies from this population have gained new traction through the U.S. culture wars and other populist media phenomena. Should climate change deniers join the mix of epistemological approaches to the Anthropocene? We have not answered such questions, which we leave to our readers to tackle.

Indeed, plenty of open questions pervade this text. Consider, for example, the difficulty in translation, to any language we know of, for the term "feral." In Romance languages, the term blends into words for wild, such as French *sauvage*. In some non-European languages, even "wild" is hard to translate. This poses serious problems for a textual analysis of the patchy Anthropocene that moves beyond the Anglosphere.

In the Meratus Mountains of South Kalimantan, Indonesia, where Anna Tsing spent several years doing ethnographic fieldwork, the Indonesian term for wild, *liar*, is used to refer to illegal and outlaw activities. It is not used to describe any feature of the natural world. Until relatively recent incursions by corporate timber companies, mining, and oil palm plantations, this was a rich, anthropogenic ecology, in which tropical forest and human livelihoods blended together easily. Wild relatives of domestic animals and crops abounded; thus, for example, chickens and pigs were able to breed with relatives in the forest. The forests were full of much-loved fruit-bearing trees, some planted by people and some occurring naturally. Plenty of foods (for example, wild cucumbers that grow only as weeds in fields but are never planted) are hard to classify in any wild-versus-domestic logic. People classified plants by their individual histories: They were either planted by people (in which case those people retain rights), or they came up by themselves (in which case common property applies). But this is not a species characteristic; plenty of kinds of plants are known in both forms. So much life would fall under the Anglophone notion of feral here.

Traditional infrastructure lent itself to this anthropogenic natural world.

Houses were built of light materials such as bamboo, bark, and thatch, which, when abandoned, decayed, supporting the regeneration of plant and animal life. House posts were often made with a kind of wood that re-sprouted even after being cut; thus, by the time its occupants were ready to move, the house itself was likely to be growing. Such houses, now despised in government standards, were easy to integrate into gardens and forests: They were full of human presence yet also had a lively and biologically diverse autonomy.

Meratus arrangements remind us: The concept of the feral offers startling insights only within a particular cultural history and political economy. Within the Anglosphere, with its heritage of British colonial rule, the human presence interrupts nature rather than joining it. Lines between wild and domestic remain sharp. Infrastructure overruns and replaces wild ecologies, creating feral effects. The Anthropocene springs up from such interventions—rather than from the human condition as a whole.

Books are necessarily written within particular cultural histories. But what does it mean to write a book in which the authors know in advance the difficulties of translation? The best we can do is to offer this awkward acknowledgment. Our aim to pluralize the Anthropocene pushes back against a mainly unselfconscious hegemony of particular habits of thought, which calcifies modes of thinking, making it difficult for alternatives to emerge. Knowing one's limitations is a first step.

TEN

Piling

What diverse processes, stakes, and concerns might become evident when different knowledge systems are brought together *on their own terms*, within and across patches? Let's call this practice *piling*. What unforeseen connections might surface? What cautionary lessons might be learned?

The patchy epistemics that we advocate—and that is set in motion through the digital exuberance of *Feral Atlas*—is a deliberately juxtaposed and often discordant approach to building knowledge that holds onto the value of place-based evidence *and* epistemological tensions in its efforts to theorize planetary processes. To practice patchy epistemics is to value the empirical and analytic traction that disciplinary specializations enable, without insisting on the imposition of hierarchies of value and truth from afar. Such knowledge practices always begin from place, and at the scale that makes sense according to the phenomena under investigation. Moreover, and as we have demonstrated throughout this book, a shared repertoire of knowledge practices incorporates the descriptive powers of drawing, poetry, music, film, sound, and more. Such forms are no longer treated as supplements to, or mere illustrations of, the "real" analytic effort.[1]

In this chapter we offer piling as a methodology for describing the patchy Anthropocene. To pile is to heap one thing on top of another without an *a priori* order: to create a structure with no foundations. Piles are unstable: They have no load-bearing capacity, and piling something too high will eventually lead to collapse. Piling is a curatorial method: a knowledge practice of gathering up and holding onto varying perspectives; a gentle and generous mode of making sense guided by the materials and insights thus brought together. Piling enacts a not-so-organized assembling of materials; it is a way of seeking out and collating with selective purpose and critical intent that resists the urge to tidy everything into neat and homogenizing bundles. It is a careful but loose sorting. This entails more than simply lumping disparate perspectives together to create a big mess. Piling has to be done with open, generative critical intent, with the curators or authors responsible for thinking with and across difference to gain new critical traction from the patchy perspectives that their piling enables.

When it comes to study of the patchy Anthropocene, we argue that piling offers an approach to building knowledge that, by working with diverse perspectives and forms, offers particular analytic strengths. This approach resonates with the character of the Anthropocene itself: a pile up of feral effects—as artist Emmy Lingscheit conveys in her 2016 lithograph *Cover the Earth* (Figure 42).

This chapter describes three kinds of piling practices: juxtaposition, displacement, and amassment. Juxtaposition is the practice of putting objects side by side. In Lingscheit's artwork, pigs are juxtaposed with bees. Displacement makes internal contradictions signify, as when escaped pigs and bees in the piece draw attention to their enclosures. Amassing means gathering together in heaps, like that of the trucks and their loads. Each of these modes of composition was practiced in *Feral Atlas* to develop and present a patchy epistemics.

Juxtaposition

As a compositional or curatorial method, juxtaposition activates the gap between things. It creates a side-by-side relational tension in which objects or ideas are positioned as both together and apart. It is an art of surprise and disconcertment—a technique of meaning making that invites its audience to participate in the backwards and forwards dynamics it creates, as they contemplate across sameness *and* difference. New and often unexpected connections become evident while distinctive differences remain in view. Importantly, too, juxtaposition does not suggest a fixed or inevitable relationship. Its power is in staging telling conjunc-

FIGURE 42. ***Cover the Earth.***
EMMY LINGSCHEIT

tures between things not necessarily—or, at least, not previously—recognized as related.

By structuring *Feral Atlas* around a series of field reports, the curators juxtaposed varied kinds of knowledge. In these reports, for example, readers are exposed to sharply contrasting ways of investigating Anthropocene patches:

Andrew Mathews: close observation. Take a look at Figure 43, a photograph of a stump: a dead Italian chestnut tree, offered by researcher Andrew Mathews.[2] You might not think there would be much to see there. But in his *Feral Atlas* report, Mathews explains how the stump tells a lively history of the forest, including its diseases, attempted cures, and worse outcomes. The tree was cut, but not killed, in the 1950s; later, new shoots grew and were cut back for firewood. All this can be seen in this stump. Even in death, trees tell stories through their shapes.

FIGURE 43. **Histories in plant form.**
ANDREW MATHEWS

Ernest Alfred: Indigenous experience and advocacy. In *Feral Atlas*, First Nations 'Namgis Chief Ernest Alfred discusses the effect of salmon pests and pathogens.[3] Discharges from commercial salmon farms, Alfred shows, pollute the coastal waters of British Columbia, threatening wild fish. Alfred's report relies on the changing experience of his community over generations; as he writes, "Long ago, my ancestors could walk across the river on the backs of the salmon."[4] His report builds, too, from his community activism, which has inspired in-depth investigation of deteriorating salmon ecologies in his area including his own collaborations with scientists.

Gillian Bogart and Gde Putra: folk songs as commentary. In their *Feral Atlas* report, Bogart and Putra introduce users to an Indonesian folk song that describes an invasive weed of wetlands, genjer (*Limnocharis flava*).[5] Folk songs can bring listeners inside local natural histories (recall Chapter 7's discussion of Lead Belly's "Boll Weevil Blues"). Bogart and Putra lead us further into political commentary.

The weed genjer was an important food for the poor, especially during times of war. The song describes the harvesting of this weed: Here is common livelihood, simple but sustaining. Yet in the 1960s the song, which had become popular with the Indonesian Left, was politicized in anti-communist government propaganda as a violent threat. Bogart and Putra discuss the hesitant return of this song today: Through appreciating weeds, people claim rights to survive.

Nils Bubandt: uncertainty—and nonsecular alternatives. Anthropologist Nils Bubandt writes about a mud volcano that erupted in East Java, Indonesia, in 2006, displacing more than 39,000 people.[6] Controversies erupted too: Was the mudflow caused by a natural earthquake or by exploratory drilling by an oil company? Instead of throwing his weight to one side or the other, Bubandt wades into the turbidity of knowledge. The most important point, for Bubandt, is indeterminacy itself; this allows him, as analyst, to move in and out of the geologists' knowledge frame. Local people explained to him that the volcano is a sacred, living being and that its vent is the mouth of a giant snake. Bubandt explains: "[G]eology—one of the most prominent sciences behind the concept of the Anthropocene—is as haunted by uncertainty as are the victims who continue to live near the mud volcano."[7]

Feral Atlas puts these knowledge practices side by side, hoping that readers will learn through this juxtaposition. Why bother? More than one kind of knowledge practice, we argue, is necessary to best describe the Anthropocene. If we care about patches, we need kinds of knowledge suited to learn about them. Ignoring local and Indigenous knowledge has often led to bad science and policy. When Canada's government gave leases for commercial salmon production in British Columbia, it was not adequately considering ecological effects—which were identified through Indigenous experience, as described by Alfred. Taking different kinds of knowledge into account can offer us better, more useful description.

Standardizing knowledge is great for particular projects, but the dream of unifying knowledge practices *for its own sake* is dangerous and misleading. A good example of the problem is the movement to promote the superiority of what some researchers have called "evidence-based" knowledge, that is, generalizations compiled from large data sets, whether reliable or otherwise. Like Humpty Dumpty in *Through the Looking-Glass*, who claimed words mean whatever he wants them to mean,[8] this movement has tried to capture the term "evidence" for a very particular set of practices, discrediting other sources of empirical knowledge as "nonscientific." The movement was born in medicine to delegitimate the insights doctors gain through their interactions with patients.[9] It has moved into conservation, where it erases place-based ecological and cultural knowledge.[10] Although this kind of "evidence" often rests on questionable data, such research struts its claim to standardize and unify knowledge, assuming that this is itself the goal. Instead, we argue that multiple knowledge practices enrich our ability to describe the world. Rather than create what science studies scholar John Law calls a "one-world world," we argue that a patchy Anthropocene needs a patchy epistemics.[11]

One important use of juxtaposition is to bridge the divide between how researchers study nonhumans versus how they study humans. Studying the Anthropocene requires both kinds of expertise—but the gap across knowledge traditions continues to block effective research. This book argues that we can do better. Our claim about the feral effects of imperial and industrial infrastructures is one suggestion; it could lead to more-than-human histories, crossing humanism and natural science.

Juxtaposition can build a bridge, that is, an opening toward creating a field in which readers are expected to be knowledgeable about *both* humans and nonhumans. *Feral Atlas* stages such juxtapositions by offering multiple field reports on the same phenomenon, each working from different knowledge traditions. An exceptionally heady group of reports was gathered to discuss the coronavirus pandemic, which was in full swing in 2020 when the atlas came to final form. This was not yet a good time for a full report; the atlas offers the pandemic in becoming. The reports are exceptionally varied. Lebanese feminist writer Lena Mounzer's "Letter from Beirut" uses intimate personal experience to document the pandemic as a mode of "living in the midst of history."[12] Figure 44 is the photograph of a 2019 demonstration; Mounzer uses it to show the political scene cut off by quarantine. She writes of how personal and political frustrations pile up in the quarantine, which itself provokes terrifying memories of Lebanon's civil war, 1975–1990. In contrast, biologist Scott

Gilbert uses the pandemic to discuss how microorganisms can cause havoc when they switch the hosts with which they are associated; coronavirus moving from bats to humans is only one example.[13] Historian Andrew Liu writes of the early spread of the virus from China to Europe through the just-in-time supply chains that have become so important to global commodity production.[14] These are just three of the seven field reports *Feral Atlas* uses to describe a moment of becoming within the COVID-19 pandemic. To understand the pandemic in the midst of history, both natural science and humanist modes of knowledge are necessary.[15]

Some juxtapositions take readers in and out of the academy, showing how vernacular knowledge can augment professional knowledge. *Feral Atlas* offers several reports on Dutch elm disease, a devastating epidemic

FIGURE 44. **In Beirut, the COVID-19 quarantine shut down political action, leaving the frustration of unfinished business. In 2019, before the pandemic, protestors demonstrated in the city's squares during the October 17 Revolution.**
RIMA RANTISI

for elms. Susan Wright narrates her experience of the elms that defined her childhood—elms that were infected, felled, and burned in the UK epidemic of Dutch elm disease that began in the 1960s.[16] This is living in the midst of history, but it is not the only history that matters in getting to know Dutch elm disease. There are other human histories, from other times and places. In one juxtaposition, the atlas offers Valencia Robin's moving poem "Dutch Elm Disease," in which dying elms accompany the grief of African American families in Milwaukee who lost sons in the U.S.–Indochina Wars.[17]

Meanwhile, the atlas offers a second juxtaposition to acknowledge the history and agility of the pathogen—the kind of story told in Chapter 9. *Feral Atlas* presents a field report by forest pathologist Clive Brasier, who describes the transformations of the pathogen as it moved back and forth between Europe and North America, on commercial wood shipments.[18] This is a riveting story of biological interactions. Brasier inspires Chapter 9's argument that this set of fungal transformations is a beacon for coming disasters, at least if global trade in living organisms continues as it has.

By placing these reports next to one another, *Feral Atlas* asks readers to consider the pandemics that affected elm trees (like the coronavirus pandemic) within their simultaneously human and nonhuman histories. If readers can get used to understanding history as multivalent—human and nonhuman—perhaps we have a chance at describing, and understanding, the Anthropocene.

One more example of linked reports in *Feral Atlas* can highlight the possibilities of dialogues between the sciences and the arts. The dialogue we advocate requires that one not limit the arts to science illustration (as some scientists might want), nor promote them as the universal human spirit (as some humanists might want). Arts, like sciences, can offer good description and place-based analysis. In this, the knowledge offered by artists joins scientists' knowledge as potential partners in juxtaposition.

Feral Atlas provides several reports on kudzu, a big-leafed vine, which, in human-disturbed ecologies in the U.S. South, tends to smother whole landscapes. The two from the arts, one from a photographer and the other from a poet, already exist in epistemic tension. Photographer Helene Schmitz gives us a series of haunting black-and- white photographs, which show the plant smothering houses, trees, and telephone poles.[19] They draw viewers out of place into the uncanny feeling of being dwarfed and subsumed by this plant. (One of Schmitz's photographs is shown in Figure 45.) In contrast, Beth Ann Fennelly's poems, "The Kudzu Chronicles," usher readers back into place with the plant's Mississippi mode of being.[20] In the fourth poem one can almost hear regional dialect:

> I asked a neighbor,   early on,
> if there was a way
> to get rid of it—
> Well, he said,
> over the kudzu fence,
> I suppose
> if you sprayed it
> with whiskey
> maybe
> the Baptists would eat it—
> then, chuckling,
> he turned
> and walked back inside his house.

Both these descriptions of kudzu are augmented by *Feral Atlas*'s juxtaposition with another report, by ecological scientists Irwin Forseth and Anne Innis, which describes the biological resources that have allowed the plant such success.[21] Kudzu, they explain, has spread across the South due to government planting programs as well as the common occurrence of abandoned cotton plantations with eroded soils. Kudzu has features, such as large leaves, that take advantage of ruined human landscapes, allowing it to take over. This ecological discussion is important context for the photography and poetry. In complement, the arts call up the affects that have made this vine so important not just in southern folklore but also in science and policy. Both are necessary for the important work of description.

Juxtaposition is just one step in building a new field of knowledge that can include both humans and nonhumans. But such building is long overdue—and getting readers to enjoy thinking across conventional lines is an important first step.[22]

FIGURE 45. ***Alabama Fields.***
HELENE SCHMITZ

Displacement

Writing informed by place-based knowledge must displace taken-for-granted generalizations built on modernist knowledge simplifications. This kind of displacement is also an act of piling: It acknowledges hegemonic truths while piling on alternatives gleaned through taking terrain seriously. Every writer of a field report in *Feral Atlas* needed this skill of displacement to draw attention to a phenomenon the writer found worth reporting. However, it is easiest to see this practice where writers are inspired by contrasts and contradictions in just how that place could be studied.

Sometimes the goal is simply to see beyond overactive generalizations. Biologist David Skelly and his co-workers took on the generalization that frogs are disappearing globally to focus their study on the particularities of New England suburbs. In fact, they found, green frogs have been doing well there.[23] Wood frogs, indeed, disappear, but not green frogs. It is a mistake to assume a single destructive trajectory. Instead, place-based research can show just which microgeographies become refuges and which species thrive. Human-made puddles in the suburbs are great for green frogs.

Two kinds of knowledge-making contrasts make this report stand out: first, the contrast between disappearing wood frogs and thriving green frogs; and second, the contrast between assumptions about amphibian decline and the situation in New England puddles. Place-based knowledge displaces stereotypes through such contrasts.

Displacement enlivens anthropologist Jacob Doherty's report about the marabou storks who have become full-year residents at the garbage dumps of Kampala, Uganda.[24] Elites and city managers hate the birds because they sit in trees in leafy residential neighborhoods, defecating. Yet, scavengers at the dumps love them, because the storks remove organic garbage from the dump, revealing the solid, long-lasting materials the people can sell. Assessments of marabou storks thus depend on class differences. By paying attention to the perspectives of both managers and scavengers, Doherty changes what can be said about storks as feral beings. Scavengers interrupt city managers' assumptions about what makes a good life in the city; they offer an alternative kind of knowledge, which Doherty makes available to readers. By piling up both kinds of knowledge, Doherty displaces the hegemony of managers' views.

These two examples—one from the natural sciences and one from the social sciences— show that displacement is not the exclusive skill of any one scholarly discipline. The arts, too, can be good at this. Consider Juliana Spahr's poem, "Gentle Now, Don't Add to Heartache," which *Feral Atlas* has classified as a field report.[25] The poem describes a stream in Ohio, near where the poet grew up. It imagines the poet's immersion in all the lives that grow in and around the stream through lists of the common names of its living beings. Some of these beings are little appreciated. A surprising number, for example, are freshwater mussels and clams, which once thrived in the cool streams of the Midwest but now have become quite rare due to stream pollution. The biodiversity that bivalves support is astounding, as is their importance for stream ecology through their work of filtering stream water. Listing names is a way of offering appreciation.

Inside the poem there is an important conjunction made, between immersion in a world of synthetics, on the one hand, and in a world of living beings, on the other:

> We let in soda cans and we let in cigarette butts and we let
> in pink tampon applicators and we let in six pack of beer
> connectors and we let in various other pieces of plastic
> that would travel through the stream.
> I didn't even say goodbye elephant ear, mountain madtom,
> butterfly, harelip sucker, white catspaw, rabbitsfoot,
> monkeyface, speckled chub, wartyback, ebonyshell, pirate
> perch, ohio pigtoe, clubshell.

The poet argues that people have let the former displace the latter, erasing the latter's existence. There has been, the poem suggests, a serious ontological shift, which makes older ways of life invisible:

> I replaced what I knew of the stream with Lifestream
> Total Cholesterol Test Packets, with Snuggle Emerald
> Stream Fabric Softener Dryer Sheets, with Tisserand
> Aromatherapy Aroma-Stream Cartridges, with Filter
> Stream Dust Tamer, and Streamzap PC Remote Control,
> Acid Stream Launcher, and Viral Data Stream.

This juxtaposition of ways of life—in effect, a displacement—is at the heart of the lament. Even in mourning, the poet draws attention to alternatives still evident through place-based knowledge:

> To sing in lament for whoever lost her elephant ear lost
>   her mountain madtom
> and whoever lost her butterfly lost her harelip sucker
> and whoever lost her white catspaw lost her rabbitsfoot
> and whoever lost her monkeyface lost her speckled chub

and whoever lost her wartyback lost her ebonyshell
and whoever lost her pirate perch lost her ohio pigtoe lost
her clubshell.

The list is a description, an act of mourning—and a glimpse of a too-often hidden alternative world.

Amassment

The multimedia digital architecture of *Feral Atlas* enabled us to pile up a certain epistemic unruliness. Because the project was not a book, the curators were freed from the fixed order or logic laid out in a table of contents. Indeed, the point of the atlas for the curators was to see what new insights and connections might emerge from what became a steadily growing stack of commissioned field reports and specialist essays over the five years it took to create.

Using the riches we gathered, the curators deliberately created new kinds of piles with which to understand the Anthropocene, and particularly the ways connections and disconnections across patches work together to produce planetary effects. *Feral Atlas* works hard to discourage readers from just reading field reports, however startling the juxtapositions and displacements. Instead, its design urges users to make their way through Anthropocene Detonators (Chapter 6) and Tippers (Chapter 4) before even arriving at a field report. Each of these analytic axes offers a form of piling. They bring users back and forth between patches and much larger spaces and times, including the planetary scale over the last five hundred years.

There is a third analytic axis in *Feral Atlas* that has not yet been explicitly discussed in *Field Guide*: Feral Qualities. Feral Qualities are modes of attunement through which nonhumans make human infrastructures affordances for their own activities. The ways that such attunements make histories was the subject of Part III. Here, however, Feral Qualities offer one more "ride-along" to explain the work of amassment, as a form of piling. (See the Appendix to learn how to find and use Feral Qualities on the website.)

Consider the Feral Quality "Industrial Stowaways," which groups feral entities that ride in industrial shipments. Thomas Bassett and Carol

Spindel's report on the European common reed (*Phragmites australis,* subspecies *australis*) in the United States explains one long-mysterious case.[26] Common reeds are native and mild-mannered members of wetland communities in North America; yet sometime in the twentieth century, they got surprisingly aggressive, crowding out other species. What had happened? After considerable detective work, botanists realized that seeds of the European subspecies had hitchhiked in the ballast materials of nineteenth-century merchant ships traveling from Europe to the United States. The new subspecies displaced native reeds and spread rapidly across sites of human disturbance, where it took over wetlands, outcompeting other plants and forming its own monocrop. The native reed had been replaced by an aggressive variant, transforming wetland ecologies.

This story seems benign when compared to a different stowaway: radiation in imported commercial foods. Kate Brown reports on the trade in wild blueberries that has blossomed in northern Ukraine and southern Belarus since the Chernobyl nuclear accident.[27] Blueberries flourish in the marshes, where they absorb Chernobyl radiation from the soil. Brown documents the local cultural and political economy that makes the blueberry trade so successful; she connects this with the international market for wild blueberries, which spreads Chernobyl radiation around the world, through human foods. "Chernobyl is on your breakfast table," she concludes.

It's both hard and easy to put the two protagonists—reeds and radiation—on the same analytic map. By grouping them in the same Feral Quality, the atlas puts them in a pile that acknowledges their movement through the infrastructures of transnational shipping. Yet this is not a comparison; it assumes radical differences that might make comparison impossible. Indeed, contrasting amassment and comparison clarifies the stakes. A formal comparison requires the analyst to standardize the two stories enough to meaningfully determine similarities and differences.[28] In contrast, an amassment lets similarities and differences fly. Both reports refer to swamps! But only one of these trades could still be stopped! Similarities and differences can fly around freely in an amassment; no standardization and objectification of the cases is necessary.

Piling the Anthropocene

By inviting multiple worldviews and embracing diverse forms of description, patchy epistemics cultivates composite perspectives. Only in this way, we argue, is it possible to adequately think with and across patches. This work requires a certain epistemic care. This is not the same as interdisciplinary translation, that is, rendering one person's knowledge legible to another. Instead, this approach requires holding open analytic spaces that recognize and respect incommensurabilities between knowledge systems—not to mention remaining attuned to the many crucial differences between Anthropocene patches themselves, which as we have argued, do not necessarily scale up, or map onto, each other.

Concerned with identifying and describing the world-making—or, rather, the world-ripping—forces of human infrastructure projects, *Feral Atlas* piles up reports in ways that help users to understand the impossibility of curating the Anthropocene within a singular planetary view. As *Feral Atlas* enables users to navigate the pile—encouraging users to dwell in the disconcerting realizations of juxtaposition, displacement, and amassment—it enacts a performative heterogeneity; it puts patchy epistemics in action, refusing structural simplicity and "smooth explanatory narratives."[29]

ELEVEN

Building and Unbuilding

FEIFEI ZHOU

"Unbuilding" is an important art for living in the Anthropocene.[1] For centuries now, elites have imagined building as the way that humans might conquer the earth. As an architect, I am constantly aware of the role of architecture in building dangerous feral effects into the Anthropocene. What kinds of building and unbuilding, this chapter asks, might challenge this?

Sometimes, we do not need to build. In winter 2021, a canal built through the Zakole Wawerskie wetland in Warsaw caused flooding in the midst of continuous heavy rain. A local activist group responded immediately by planning a small dam to divert the flood. After three days of preparing sandbags, wheelbarrows, and other tools for the construction, they arrived at the flooding site only to discover that beavers had built a dam already. The beavers were way ahead of the humans in managing the flood.

I learned this story from curator Gilly Karjevsky when she and I were planning a workshop concerning forms of noticing and preserving this beautiful and complex wetland. Instead of advocating any active intervention, our proposal offered ways to collectively pay more careful attention to nonhuman practices and stories.

Although the workshop never happened, the curator's account stuck with me in thinking about the concept of unbuilding as a method of "collaborative survival."[2] In this chapter, "unbuilding" has multiple meanings. First: holding back. In the case of beavers, humans simply did not need to do anything; our nonhuman companions offered effective and efficient engineering solutions that made human interventions seem redundant. Second: challenging assumptions. Though caught by surprise, the humans in this story did not disturb the beaver dam and simply welcomed the natural mitigation. This example of the beavers' dam manifests an unbuilding of anthropocentric perspectives and ideologies. Unbuilding embraces collaborative negotiations in the building process. Third: removal. Unbuilding can require taking out projects that have caused disruptive effects. Unbuilding in this sense allows ecologies and cultures to recover. The current dam removal movement in the United States, for example, has brought back wildlife, including beaver communities, that were previously endangered by the feral effects of water engineering projects. (This chapter will have more to say about the dam removal movement later on.)

Unbuilding works together with building. Sometimes, we need to build the foundations for unbuilding. This was the case for *Feral Atlas*. One of our challenges was *building* a collaborative project to support designs for unbuilding. Collaboration often means letting go of preconceptions, that is, unbuilding. The first part of this chapter describes how the participants in the *Feral Atlas* project worked across our differences in building the website by negotiating and unlearning parts of our discipline-bound views and understandings. Building and unbuilding worked together in this process. The second part of the chapter discusses how collage as a methodology worked to make room for diverse knowledge-making practices in the construction of *Feral Atlas*'s Anthropocene Detonator landscapes. Unbuilding is crucial for collaborative building. The last section of the chapter reevaluates the ethos of the discipline of architecture through examining two modes of unbuilding as architectural practice. Architects only properly acknowledge the dangers of the Anthropocene when they begin to incorporate unbuilding into their thinking and practices.

Unbuilding Preconceptions: How Precision Drawing Can Open Up Mappings

Any collaboration is hard. Transdisciplinary collaboration can be particularly challenging, as it requires unlearning disciplinary knowledge, ideologies, methodologies, approaches, habits, languages, styles, and so on. It's hard. A transdisciplinary, multi-genre, concept-stretching collaboration of more than a hundred people can seem almost impossible.

But we made it happen.

In making *Feral Atlas*, the editorial team had to work closely with other designers, seeking out common ground across our disciplinary and personal differences and thriving through "productive friction." This kind of friction represents the results of encounters in which varied epistemologies collide and coexist.[3] However, through collaborative efforts in acknowledging, understanding, and attuning to these tensions within differences (such as conversations, debates, bonding, and simply listening), we were able to create a digital site that embraces diverse modes of knowledge-making practices. This would not have been possible without productive fiction.

One important moment, which I think of as an instance of "mapping through unmapping," sticks in my mind. The team's first collective exploration of what the atlas "maps" might look like took place in a workshop at the Moesgaard Museum in Aarhus, Denmark. The editors met with digital designers and builders, along with my architect colleague Matthew Darmour-Paul, and we began exploring how we might visually present some of the key concepts in the atlas in unconventional ways. In Part I of *Field Guide,* we introduced how we stretch the concept of maps beyond conventions—shifting scales, diversifying genres, paying particular attention to feral dynamics through motions and transformations. But this idea would not have come into being at all without the productive friction of our initial conversations. When presented with the concept of diversifying map-making practices, some of the team members were not convinced. Can a visual representation that differs vastly from satellite imagery even be called a map? Digital designers on the team who were most familiar with Google Maps—data-based, relatively accurate to geography, and easy to read—initially found this idea unsettling.

Satellite technology introduced modern humans to an entirely different way of viewing the planet: an elevated and distanced point of view that requires no personal observation. Modern cosmopolitans have come to believe that this is the only accurate map; all other mapping technologies, we think, are archaic. At the workshop, advocates of diverse mapping techniques had to think: How could we introduce alternative representations as maps?

One technique was show and tell. Precision drawing of the kind perfected by architects turned out to really matter in convincing participants that there was more to mapping than data visualization. Some workshop participants used the compelling images in the book *Petrochemical America* to show what might be possible.[4] The book consists of haunting photographs of Louisiana's petrochemical landscape taken by Richard Misrach; landscape architect Kate Orff unpacks the photographs with diagrams. There is more here than what can be seen in a conventional map. Figure 46 offers an example: The misty, gloomy, and barren landscape, captured by Misrach (on the righthand side of the image), was once the Morrisonville Settlement in Louisiana, where generations of African Americans built their communities from scratch. Despite an environment marked by poverty and a lack of civil rights, these houses, depicted in Orff's drawings, were privately owned spaces where Black lives and freedoms could be valued and celebrated. In the 1950s, Dow Chemical Corporation began its infrastructural expansion into the area, and eventually bought out almost the entire community for radical industrialization. All the houses were demolished, and residents were displaced, leaving no visible traces of their previous lives. The "mapping" practice here captures the life of the past, the "ghostscapes" of the present, and the connections between the two.[5] The vividness of such an example helped us as *Feral Atlas* editors overcome anxieties about forming new kinds of maps.

Another example suggestively modeled by architects came later: maps that engage viewers with the state of the soil. A map of human-disturbed soils was created by historian Frédérique Aït-Touati and architects Alexandra Arènes and Axelle Grégoire (the *Terra Forma* group) to show the remaking of soils through the interaction of human and non-human building projects.[6] The mapmakers introduce an alternative point of view on reality: Instead of representing earth as the familiar blue marble globe

FIGURE 46. ***Community Remains, Former Morrisonville Settlement, Dow Chemical Corporation, Plaquemine, Louisiana.*** **1998.**
KATE ORFF AND RICHARD MISRACH

(the view from space), their soil map unfolds from the center. The atmosphere is at the center of the diagram; the soil sits below it, inside out, expanding as we look deeper into the ground, that is, outward on the map (Figure 47). By turning the earth inside out, the map encourages architects to extend our interpretation of the built environment beyond what is visible above ground. Excavation, extraction, discharge, and reclamation are all part of the built environment. The above-and-below-ground is an active, interconnected, and entangled system of being.

At the *Feral Atlas* workshop on maps, the detailed representations of architects' perspectives—however divergent and strange—were powerful in convincing skeptics. My own work was part of it: a "plan" perspective is one kind of map, I showed; a journey itinerary is another. How could drawings made with so much care be completely wrong? It wasn't the argument per se; it was the series of visuals that brought us together. In the energy of this wind, which swept doubters into the conversation, we hung a range of spatial representations across the walls, challenging ourselves to consider what the best mapping practice for a particular feral phenomenon might be. We played with each other: Might a scrunched-up paper

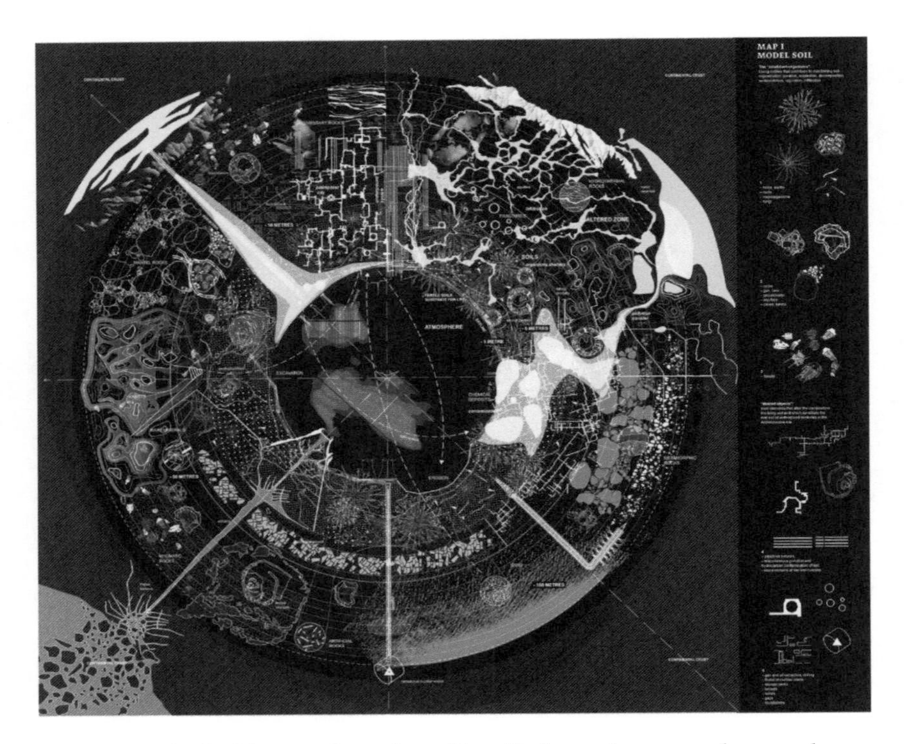

FIGURE 47. **An inside-out view of earth's soil places the atmosphere at the center to emphasize planetary containment; the depths show anthropogenic interactions.**
FRÉDÉRIQUE AÏT-TOUATI, ALEXANDRA ARÈNES, AND AXELLE GRÉGOIRE

map still be a map? Was the view from the window a map? It took a lot of unbuilding to get to the practice of building.

Collaborating Through Collaging: Making Room for Others

Collage is a reassemblage of forms that together create a new artwork. To create landscapes that showed the sweep and diversity of continent-crossing historical ruptures in building programs, which our team called Anthropocene Detonators, I decided to use collage. Because it is built through diversity, collage could help us bring BIPOC artists and artists from the Global South into the process of building an Anthropocene story. Collage, I hoped, might make room for multiple modes of storytelling.

I created four Anthropocene Detonator landscapes to tell the story of acceleration, capital, empire, and invasion, respectively. Each landscape is drawn from a different perspective, to comment on the key modes of changing the earth associated with that world-rocking set of transformations. For only one of these landscapes did I go out of my way to present a single perspective by itself: *Capital*. Here the homogenization of the world into a grid of commodity production is the point of the landscape. It, too, is the only drawing that did not include collaborators. For the others, I embraced the patchy epistemics this book introduces; I worked to make my drawings offer a constant shift in stories, places, and ways of seeing. Together, they tell of major shifts in world-building dynamics. But they do this through unhomogenized patches.

As mentioned in the Introduction of *Field Guide*, storytelling is a crucial way of noticing and representing. The Anthropocene Detonator landscapes further extend the nature of storytelling by visual representations, and through multiple hands, perspectives, styles, and narratives. Drawing is powerful. Humans have long been using illustrations to archive and communicate. The contemporary Chinese written language is developed from ancient hieroglyphs, a continuation of the historical logographic system of storytelling. Drawings have the power of language in unfolding complex, intertwined, and sometimes hidden more-than-human histories.

But to pluralize the Anthropocene through visual narratives, the power of diverse modes of drawing practices needs to be brought to the center. As Part III of this book argues, an important step of decolonizing natural history is to recognize other ways of telling histories. The Ngurrara Canvas, for instance, challenges the exclusivity of the text-based legal system in Australia through the power of collaborative art.[7] To prove that Aboriginal law has long existed and been practiced at the Great Sandy Desert of Australia, more than sixty artists from different Indigenous nations came together to create the Ngurrara Canvas, based on collective law and experience. Where the dominant legal system determines land ownership through written words, based on the English language and concepts, Aboriginal law and custom are commonly affirmed through oral transmission and shared understandings.[8] Moreover, the settler government struggles to recognize collective ownership, instead emphasizing individual rights. Aboriginal artists determined to tell their histories in a different way—in their own way. Before the Ngurrara Canvas became one giant drawing, the artists began by each painting a different part of the canvas.[9] They soon realized that individual pieces could not represent their common knowledge and deep-rooted connection with their land as one group.[10] Working together across generations, communities and language groups, they eventually produced a unique painting that let "all the right people to tell their stories and paint their own country."[11] The canvas was presented as the key evidence in the 1997 Ngurrara Native Title Tribunal. This collective painting practice challenged the colonial legacy of land ownership determinations by embodying other kinds of history telling. In *Feral Atlas*, we had a chance to make our own experiments with collective storytelling in the Anthropocene Detonator landscapes.

Acceleration

The most commonly known form of collage is perhaps the assemblage of cut-and-pasted prints or objects to create a new image. When each element is deconstructed from its original form, medium, and subject matter, a new context emerges through intentional juxtapositions. I find collage to be an exciting method with which to reimagine existing forms and suggest sometimes surprising new meanings.

The Detonator landscape for *Acceleration* embraces this multimedia juxtaposition by embedding photographic cutouts of a sculpture of the manananggal, a mythological monster, created by the Filipinx artist duo Amy Lien and Enzo Camacho. The manananggal is a woman who is split at the waist, her head and torso flying through the air while her legs and pelvis remain earthbound. Her top half wails and flies, trailing her intestines; she drinks blood and sucks at the viscera of sleepers. To tell of this disruptive, unstoppable hunger, the artists created five sculptures of manananggal, with ten part-bodies scattered around the city of Berlin (Figures 48 and 49). The monsters are a reminder, a catalyst, according to Lien and Camacho, exemplifying the urgency of dissatisfaction.

The figure of the manananggal is given new symbolic implications in the world of *Acceleration*. Her top half flies above a Philippine sugarcane plantation, haunting this colonial crop. Meanwhile, her legs are tangled up in a knot of marine plastic. Her sucking tongue moves back into the

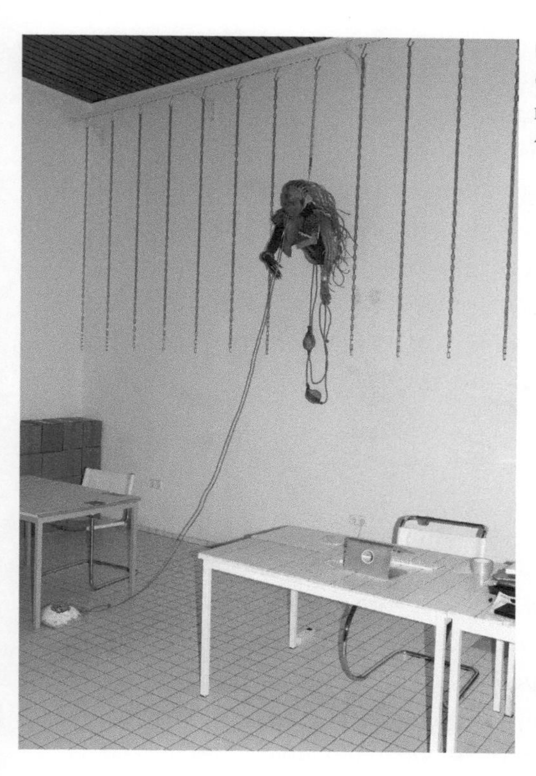

FIGURE 48. ***WAKA WAKA GUDETAMANANGGAL.***

INSTALLATION AND PHOTOGRAPH BY AMY LIEN AND ENZO CAMACHO

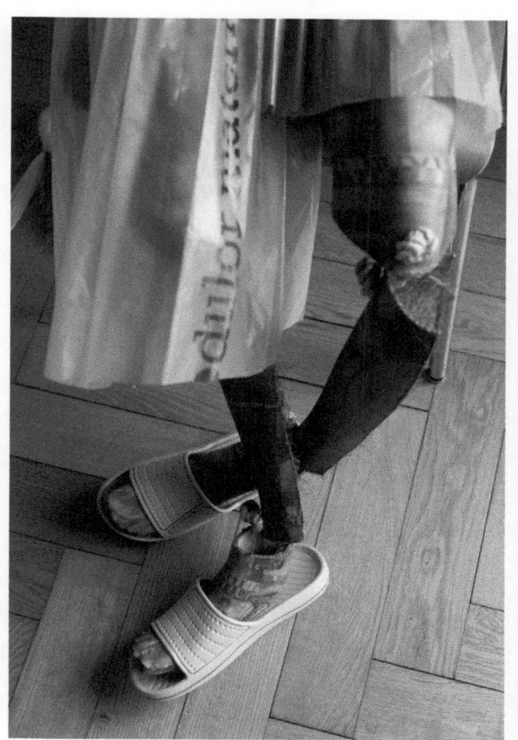

FIGURE 49. ***BRUTTO NETTO MANANANGGAL.***

INSTALLATION AND PHOTOGRAPH BY AMY LIEN AND ENZO CAMACHO

entangled fishnets: Even as she haunts the countryside, her hunger has turned to the suffocating clot of more-than-colonial, more-than-rural waste. She seems uncontained but is also stuck.

"Her split body is a crack in our community," Lien and Camacho suggest of the metaphor behind the manananggal's split body in her original context.[12] In *Acceleration*, she, too, represents a crack; driven by relentless production, consumerism, and exploitation, she opens up disjunctures in the contemporary world. Even mythical figures are stuck in toxic residue.

Empire

Ghanaian British Larry Botchway's contribution to the work of collaging *Empire* is a commission-based collaboration involving numerous conversations and revisions. Being familiar with Botchway's previous research on oil palm plantations, the editors invited him to create an illustration for the *Empire* landscape that would elucidate contemporary legacies of imperial infrastructure, particularly in the context of Africa. The positioning of his insert on the *Empire* landscape is intentional (Figure 50). Directly above his drawing, readers can see an aerial diagram of the Cape Coast Castle in Ghana, an infamous slave fort where European traders imprisoned kidnapped African people and held them as colonial wares for trading. With its "door of no return," it was the miserable last stop for enslaved Africans before being displaced to America or the Caribbean. (Notice the drawing of the *Brookes* ship sailing towards the hybrid island with features from the Greater Caribbean.) Imperial power was also fueled by the stolen raw materials such as gold and ivory. Parallel to Cape Coast Castle lies the heavily guarded trading port, where stacks of timber, ivory, and other goods are waiting to be shipped afar. Below, the entire terrain is covered by oil palm plantations, leaving no trace of what grew there before.

Botchway's contribution elucidates Ghana's current conflict between exploited local laborers and distant agro-industrial investment (Figure 51). The Ghana Oil Palm Development Company (GOPDC), a Belgium-based company, controls approximately 20,500 hectares of oil palm plantation, which they purchased through the state. The land was then leased back to local farmers for industrial oil palm plantations, in return

FIGURE 50. **Detail of *Empire* (5).**
FEIFEI ZHOU WITH LARRY BOTCHWAY

for their entire harvest as a form of repayment. As agricultural land is fully taken over by monocrops, local farmers, without land and property ownership, are left with little food and economic security.[13]

Botchway illustrated the local farmers' despair and resistance to this form of contract-based agriculture practice through the depiction of people—from across historical eras—tearing up the contract. This contract visually merges into the industrial grid of the oil palms, effectively depicting the transformation of the once diverse landscape into a monoculture (Figure 51 bottom left). Botchway's insert exemplifies the penetration of neocolonial control in today's global political ecology. Nature becomes the pawn of imperial capital investment and extraction.

FIGURE 51. ***Contract.***

LARRY BOTCHWAY

Invasion

"All Aboriginal art is political," says Noongar writer Claire G. Coleman, writing about art both as a material record of Aboriginal people's shared love of their homeland and as a rebellion against colonization and exploitation.[14] Like the Ngurrara Canvas, landscapes in Aboriginal paintings are more than aesthetic subject matter. They are at once artistic renderings and material evidence of livelihood practices and land ownership of their country prior to European settlement and continuing within it.

The *Invasion* Detonator landscape in *Feral Atlas*, viewed on screen through a continuous horizonal scroll, depicts European settlers' fantasy of a linear progression through land-grabbing and reforming. To return to Le Guin's "Carrier Bag Theory," this is a spear-like mentality for storytelling.[15] But here the fantasy of settler progress is countered by numerous "gathering" places and moments, as these not only document the invasion but also show persistence, resistance, and struggle from Native communities.

Personal experiences matter in explaining the violence and exploitation that has continued its legacy in today's systematic inequality and racism. Nancy McDinny's painting *Story of Mayawagu,* collaged in the *Invasion* landscape, captures the feisty refusals of her great-grandfather Mayawagu, standing up to the white settler military police who came to capture him (Figure 52). Completed in 2013, it belongs to a series of paintings McDinny created to challenge the way Australian history is told—by portraying the standpoint of those who have been through it. The invasion of settler colonialism involved not only the massacre of Aboriginal people, but also the erasure of their history of environmental kinship by the renaming of hills, rivers, and ridges. Mining, fracking, and other infrastructural projects, too, are invading their ancestral lands and resources as a continuation of the colonial past. Aboriginal people have not stopped fighting to protect the sacred land that is their home.

The *Invasion* Detonator landscape can be read from right to left. That linear sequence shows the gradual penetration of invasive settler colonialism as it violently altered existing more-than-human relationships. Inspired by McDinny's work as activist and artist, *Feral Atlas* editor Jennifer Deger flew to Darwin to meet with her and explain our

FIGURE 52. ***Story of Mayawagu.***
NANCY McDINNY

project. Rolling out an early draft of the *Invasion* landscape, she asked McDinny if she would be interested in allowing us to incorporate her painting into the broader *Invasion* landscape. Not only was McDinny delighted to license her image to *Feral Atlas*, she had no qualms about the original rectangular painting being cropped and collaged into this wider perspective by non-Indigenous artists. Indeed, it made complete sense to her that her painting become part of a broader story about invasion and the intergenerational struggles for environmental and social justice that it continues to generate.

Another collaged drawing on the *Invasion* landscape is *Heritage* by First Nation Canadian artist Andy Everson, whose giclée print depicts the melting Comox Glacier that is culturally and ecologically significant for the K'omoks First Nation. To decolonize natural history, it's crucial that Indigenous perspectives, and the counternarratives they offer, claim a central place. In this instance, that meant ensuring that as editors we held space for Indigenous stories, told in the ways that Indigenous people have chosen to present them.

Making of the World: Preservation Through Unbuilding

Buildings can radically modify surrounding ecologies in unintentional ways, but most of these effects go unnoticed, especially by those who design them. Look around: We are at a critical point where we can no longer turn a blind eye to the long-term social and ecological consequences of the construction industry. The globalization of capital makes material access and extraction easier than ever, but these sleek systems simultaneously make it easy for most consumers to remain oblivious to feral effects, including ongoing practices of violence and dispossession. They distort the timeline of ecosystems by accelerating growth and shortening the lifespan of many nonhuman beings, stressing ecosystems beyond their tipping points.

The examples below introduce the possibility of "unbuilding" as a building strategy to restore a more ecologically attuned timeline. Unbuilding, here, will be explained in two senses: first, removing, in the case of the dam removal movement currently taking place in the United States, and second, building nothing.

Unbuilding as Removal

There has been increasing advocacy for dam removals in the United States, particularly since 2021, to restore local ecosystems and improve public health. On November 17, 2022, the largest dam removal in U.S. history—of four dams from the Klamath River—was finally approved by the federal government after decades of advocacy by Native American tribes in the Klamath River basin in California and Oregon.[16] Not only does this unprecedented act mark a victory for water rights for Indigenous communities, but it also shows how unbuilding large-scale infrastructures can become an integral and necessary part of ecological and cultural restoration.

To understand why the unbuilding of the dams is important, let's start by understanding the detrimental effects of dams. Dams interrupt the flow of the river, inhibiting the lives of all the creatures that depend on this flow, from migrating fish to the plants, bivalves, and insects that depend on the pulse of flooding and retreat. This is an example of system rupture, as discussed in Chapter 4.

Consider salmon, one of the creatures most adversely affected by dam construction. Before the Klamath River dams were built, the rivers of the Pacific Northwest had the pleasure of the abundant company of this economically and symbolically important migrating fish. For the local Karuk, Yurok, Hoopa, and other tribes, Chinook and Coho salmon have been both a primary food source and a spiritual symbol.

Dams create abrupt impediments to salmon's essential migration route. Salmon lay their eggs in freshwater, and the eggs hatch and develop into juvenile fish. When they reach maturity, they swim to the ocean for their adult life. Once ready for breeding, they then return to upstream spawning grounds for their reproduction. Without this vital commute twice in their lifetime, salmon cannot complete their life cycle.

Dams not only block migration, but they also slow currents and obscure the work of tides, in turn causing other harmful feral effects: The smooth, leveled surfaces of the dams' reservoirs become breeding grounds for toxic substances such as cyanobacteria. Also known as blue-green algae, these organisms form colonies on the water's surface and deplete oxygen, spreading toxicity. This means that the majority of the small number of salmon who do manage to return to their spawning grounds become ill or suffocate to death. Chinook salmon, for instance, saw a decline of over 90 percent in numbers over the course of the twentieth century, resulting in their being listed as an endangered species under the California Endangered Species Act in 2021.[17] A birthplace for the salmon has now become a burial ground.

PacifiCorp, the power company behind the Klamath River dams, has attempted to rely on "building" strategies, such as upgrading aging dams with fish ladders, fish screens, and other efforts for improving poor water quality that were not part of the initial design consideration. But these efforts have come with a huge cost, as much as hundreds of millions of dollars as a perpetual investment. Unbuilding dam projects makes sense not only ecologically but also economically.

Although the removal of the Klamath dams is set to begin in 2023, many, including Native American tribes and environmentalists, can't wait to see the ecological revival that free-flow streams will bring. The Penobscot River Restoration in Maine, a preceding dam removal project that the Klamath River Renewal was modeled after, saw promising results on multiple fronts.[18] Studies show that Atlantic salmon migration has returned to its free-flowing–river levels, and nearly a thousand miles of aquatic habitat has been renewed for eleven species of native sea-run fish.[19] Steady water flows brought back good water quality, as well as a diversity of native plants. At Jamawissa Creek in New York's Hudson River Valley Oscawana Park, riverkeepers and volunteers have been wonderfully surprised by the presence of multispecies travelers.[20] River creatures such as mussels, fish, river herrings, and eels are migrating up from the Hudson River estuary without obstruction, thanks to the recent removal of a 100-year-old dam.[21] Across the United States, a record 2,025 dams have been demolished as of February 2023, with 67 removals in 2022 alone.[22] A total of $2.4 billion for dam removal and restoration was approved and recognized by the Infrastructure Investment and Jobs Act in 2021—an indication that free-flowing rivers are now recognized as infrastructure and that sustaining more-than-human lives is both a matter of survival and of the preservation of cultural heritage.[23] This could not be realized without unbuilding, and the complex political work required to achieve it.

Unbuilding as Not Building

"The work of an architect is not only to build. The first [thing] to do is to think, and only after that are you able to say whether you should build or not."[24] When discussing their project for Place Léon Aucoc, a small square in Bordeaux, France, Anna Lacaton explained the philosophical intention behind her and her partner Jean-Philippe Vassal's design decision. In 1996, the architect duo Lacaton and Vassal were commissioned to redesign the public square within a larger beautification plan for public spaces in the city. Confronted with the condition of the existing site—tall lime trees, benches, well-designed facades on surrounding public housing—the architects paused the design process at the thinking phase. They spent time observing the daily life occurring on the site, conducting interviews with local inhabitants, and the conclusion became obvious: Any changes would be unnecessary. As they explained in their project statement, "Embellishment has no place here. Quality, charm, life exist. The square is already beautiful."[25] Doing nothing was their design decision. As British architect and design critic Oliver Wainwright stated, "Sometimes the answer is to do nothing."[26]

Lacaton and Vassal's philosophy of building nothing offers a daring and critical argument for us: Think (hard) before building. To understand the critical importance of unbuilding, particularly in the context of our current environmental crisis, we must return to the discussion of feral effects once again. The built environment easily contributes around 40 percent of carbon emissions. What is left out of that calculation is the unmeasurable, unnoticed, and hard-to-get-rid-of feral effects that are unintentionally brought about by the construction industry. These effects devastate ecologies, in ways both small and large. Offshore infrastructures create artificial cantilevers and vertical surfaces that facilitate fast reproduction of jellyfish, creating jellyfish invasions.[27] Industrial nurseries gather exotic plants and form monoculture conditions that render plants across the world vulnerable to fungal and bacterial infections, causing plant epidemics.[28] These feral effects are on top of the carbon emissions discussion; and they cause planetary catastrophes. Feral effects of the building industry are not part of the discourse yet, but they should be.

Some of those searching for solutions believe that active building through advanced technology can solve or ease the environmental impacts of the building industry. Techno-heavy concepts such as smart farming emerge as remedies for ecosystem breakdown, food shortages, and labor optimization. Strategies such as carbon offsetting have been employed by large corporations for mitigating their carbon footprint by carrying out ecological restoration projects elsewhere. But "solutions" like these have not proven that they can satisfactorily address the larger issue, and they exacerbate other feral effects. Less-than-flavorful smart-farm vegetables grown in a lab where they are crowded together within a meticulously controlled environment have little immunity to disease infestations or technical failure. Carbon offsetting not only turns pollution and extraction into an acceptable practice for rich businesses but can also overwhelm local ecologies by negligently introducing new species. In one city in Indonesia, mangroves were planted along the shoreline as a strategy to both mitigate flooding and compensate for the deforestation of mangrove swamps for urban developments, but they were planted immediately in front of a seawall. As tides struck back after clashing into the concrete seawall, the mangroves were knocked over. They could not survive the constant trauma. Active building programs will only work if we can provide an environment for nonhuman species to care for themselves.[29]

A Last Cautionary Tale

In this context, it's necessary to bring unbuilding into the contemporary discourse of the built environment. Let's consider one last failure: a case of how active building with the intention of conservation and renovation has, in fact, destroyed a marginalized but otherwise quite successful historical community. It encourages us to critically reevaluate the socioeconomic forces behind building projects to understand what other kinds of value have been overlooked.

On February 14, 2021, a devastating fire nearly wiped out a 400-year-old village in Yunnan Province, China. Wengding Village, of the Wa people, is located close to the China–Myanmar border. As part of a larger historical and cultural territory, previously known as the Wa state, Wengding Village arguably remains one of the best-preserved Wa ethnic cultural regions (Figure 53). In 2015 an article in *Chinese National Geography* referred to Wengding Village as "the last remaining primitive tribe," a slogan that was quickly adopted by the mainstream media for ecotourism advertisements.[30] The new tourist interest drastically transformed the village's cultural and architectural dynamics.

The vernacular architecture of Wengding Village featured an indoor fire pit that was constantly lit, providing warmth, light, and resources for cooking. The typical houses were stilt-structured and made of locally sourced materials such as bamboo, timber, and thatched roofing. These are, indeed, very flammable materials. However, the Wa people, who have lived with indoor fire pits for thousands of years, maintained their ancestral knowledge to cohabit spaces closely with fire, which is sacred for their cultural identity.

These dynamics shifted when investments boomed for the tourist industry, attracting many investors without much interest in the cultural significance of Wa vernacular architecture. Developers saw the existing

FIGURE 53. **The village of Wengding, in China's Yunnan Province, before the devastating fire in 2021.**
ZHIQIANG XIAO

houses as obsolete and underdeveloped, particularly the "dangerous" indoor fire pits, which horrified them. Why would any tourists want to "play with fire" in such flammable and outdated buildings? The government decision came quickly: A new village was constructed to host the villagers, thus "preserving" the vernacular architecture and modernizing the Wa community, despite local objections.[31] The new, modern village was designed with a unifying architectural style: brick and concrete houses with red porcelain roof tiles and off-white walls.

By 2017, the vast majority of Wa villagers had moved to the new village, allowing the old village to be adapted merely for tourism. Safety measures around fire were implemented, including removing the indoor fire pits and moving them outdoors. The old village, once a prosperous cultural habitat, had become a decorative shell of tourist attraction, lacking the vital care of the Wa people through daily maintenance and performance of domestic tasks.

When an unattended outdoor fire burst out of control, almost the entire historic site was obliterated, leaving only four homes standing. Although the official cause of the fire remains unknown, one thing we do know is that Wengding Village was safe from burning before building was deemed necessary.

TWELVE

Beyond Piling

JENNIFER DEGER

with

PAUL GURRUMURUWUY,
ENID GURUŊULMIWUY,
WARREN BALPATJI,
MEREDITH BALANYDJARRK,

and

VICTORIA BASKIN COFFEY

To curate is to activate an oscillating field of relationship: to orchestrate spaces of betweenness with interpretive purpose and revelatory intent. If that sounds stodgy, the key for those of us who seek to learn with and through curation is to hold onto a sense of play, remaining wary of any urge to pin things down once and for all. Let the world reveal itself on unexpected terms. Open up to other forms, other voices, and the new conjunctions they can enable. Find the insights—and the ethics—in that.

In *Field Guide,* we've made the case for piling as a curatorial method for working with, and across, different knowledge traditions. We've shown how, by bringing together diverse forms of knowledge, an inclusive heterogeneity is enabled that can both expand and particularize appreciations of what is happening as Anthropocene worlds rupture and reconfigure. We've described how *Feral Atlas*'s digital architecture puts these patchy epistemics into action, creating an aesthetics of both dissonance and unexpected connection, thereby disrupting persistent academic tendencies towards comparative hierarchies and disciplinary silos. But what about other ways for researchers to respectfully bring differences into relationship? Are there methods beyond piling? Other modes for assembling a patchy epistemics?

This chapter describes how *dhä-manapanmirr* guides the creative collaborations of Miyarrka Media, an arts collective based on Aboriginal lands in northern Australia, with which I have worked for the past fifteen years under the guidance and leadership of Yolŋu elder Paul Gurrumuruwuy.[1] In other writing I've used the term "curation" to describe our shared efforts of gathering, ordering, assembling, and juxtaposing our various points of views—both textual and visual—into a single work of description and analysis.[2] Lately, however, we've begun to work directly and explicitly with the foundational concepts that shape Gurrumuruwuy's commitments to co-creativity. This is making an exciting difference.

For the purposes of *Field Guide*, I write in dialogue with images made by Miyarrka Media to offer a glimpse of a work-in-progress experiment in using digital media to share Yolŋu ecological values with global audiences. Although this chapter is short, it nonetheless deals with complex ideas, concerns, and histories. I am hopeful that by using a selection of video stills from this new project, I'll be able to at least gesture towards the distinctively inclusive social aesthetics that an intercultural research collaboration attuned to the generative potential of *dhä-manapanmirr* can set in motion.[3]

Simply translated, *dhä-manapanmirr* means bringing together, moving together, or connecting. But it's the manner of this coming together—and what happens in the process—that matters most. Yolŋu presume that when gathering as a team with an agreed-upon intent or purpose, whether in a ritual or a digital media project, each participant offers their skills, histories, and ways of seeing in the service of a common goal.

Dhä-manapanmirr provides both the method and the ethos through which our intercultural arts collective attempts to bring once distinct and separate worlds into relationship. Miyarrka Media makes exhibitions, films, digital artworks, and books that explicitly reach across cultures and generations. Through Yolŋu-led experiments with voice and form, our team of Aboriginal and non-Aboriginal researchers model ways of thinking with sameness and difference inspired by the "both-ways" relational imperatives that enliven so many Yolŋu intercultural interventions and initiatives.[4] As we work together to arrange images, sounds, voices, and words—whether in a gallery, on a screen, or within the pages of a book—we don't seek to simply record, archive, or explain Yolŋu life. Rather, we curate experiential zones of affect and critical reflection, seeking to create the possibility for new zones of overlapping meaning, purpose, and mutuality to emerge for diverse Yolŋu and balanda (European, white, or non-Aboriginal) audiences—without downplaying or otherwise diminishing the differences that define us.

Yuṯa anthropology insists that a relational dynamic of sameness *and* difference provides the necessary, and mutually enlivening, grounds for coming together and coming to knowledge, whether on the ceremony ground or in response to the new challenges and demands of settler colonialism, social media, and global capitalism. Here juxtaposition is always carefully enacted, under Yolŋu leadership, with a specific social intent. The Yolŋu principles that ground this work insist the imaginary "we" invoked by this book is not utopian, but a foundational and yet always dynamic fact of life that must be recognized, activated, and nurtured. My Yolŋu collaborators insist that such constellations of knowledge and mutual regard are not only necessary but achievable (the last century of colonization of north Australia notwithstanding), if you orientate yourself to the world with an eye that identifies, affirms, and amplifies instances where sameness and difference can be held as mutually constitutive.

This creative vision is shaped by the extraordinary and well-documented Yolŋu talent for recognizing and affirming the possibility of forging meaningful and reciprocal relationships across all manner of what appear, at least to many balanda, to be foundationally incommensurate differences, including those of race, culture, and species. This is finessed—by being made moving and particular—by Gurrumuruwuy and his family, whose energetic and unshakable belief in the world-making force of the distinctly, but not exclusively, Yolŋu art of connection animates all that we do.

This is how we introduce these themes in our book *Phone & Spear*:

> Enid Guruŋulmiwuy: Balanda look at the world and see a person there and a tree over there and they see them as separate. That's so boring!

Paul Gurrumuruwuy: This book is about reconciliation. But our way is different to marching in the street.

Jennifer Deger: Let's start by affirming the pleasures of play, association, and canny recognition: the way the world can be reconfigured and made anew through deliberate acts of combination and recombination.[5]

For all its bright colors and playful collage, *Phone & Spear* is a serious and considered work. It describes worlds intensely reconfigured by mobile phones, showing some of the gloriously inventive ways that Yolŋu use their ringtones and phone-made artworks to amplify the call and reach of Country. Working within a Yolŋu economy of knowledge, it conceals even as it reveals. Anything deemed sacred knowledge, or deep knowledge, is not for publication, and certainly not meant for balanda, so that the book is as much shaped by the things not said, shown, or even gestured to as it is by those made explicit. Yet for Yolŋu, those depths of knowledge are there, for those who know how to look, how to see deeper layers and connections.

You may recognize this cycad (Figure 54) from the Introduction to *Field Guide*. Here it is again, but as originally presented in our book with a patterned team portrait of human collaborators collaged alongside. This juxtaposition emphasizes Gurrumuruwuy's deep connections to a specific constellation of kin and country, and by extension, his team's authority to speak with and from the structuring foundations of ancestral lands and connective knowledge. In the context of our book the two images combine to form a single family portrait.

In our new project we are focusing more directly on the expressive power of Country. Our aim is to communicate a living environmental ethics from the sandy shores of the outstation for which Gurrumuruwuy is custodian. As beaches become sites of increasing ecological vulnerability across the planet, Gurrumuruwuy is convinced that the place in which he lives has messages of global relevance. The video stills that we share here are part of this larger project of creation, collection, and curation for a website that Gurrumuruwuy envisages will deliver a "gift of connection and care" from his beach to other beaches.

As we slowly record, select, assemble, design, and animate sounds, images, and texts for this website, we are again curating an experiential

FIGURE 54. **Miyarrka Media team portrait, 2019.**
MIYARRKA MEDIA

connecting of worlds. While bearing in mind the fleeting allure and ambiguous intimacies that digital spaces can provide, we experiment with ways of making the voices of sea Country perceptible to audiences who will never have the pleasure of directly sitting in the sand and yarning with Gurrumuruwuy and his family while gazing out to sea.

As we foreground and amplify the voices of nonhuman kin, we seek to make a more-than-human ethics of care palpably manifest through the movements of sand and saltwater across the screen. Our aim is to enable the beach itself to demonstrate *dhä-manapanmirr* as a more-than-human imperative essential to the ongoing-ness of life itself. Yolŋu call this way of showing oneself and one's values *milkunhamirr.*

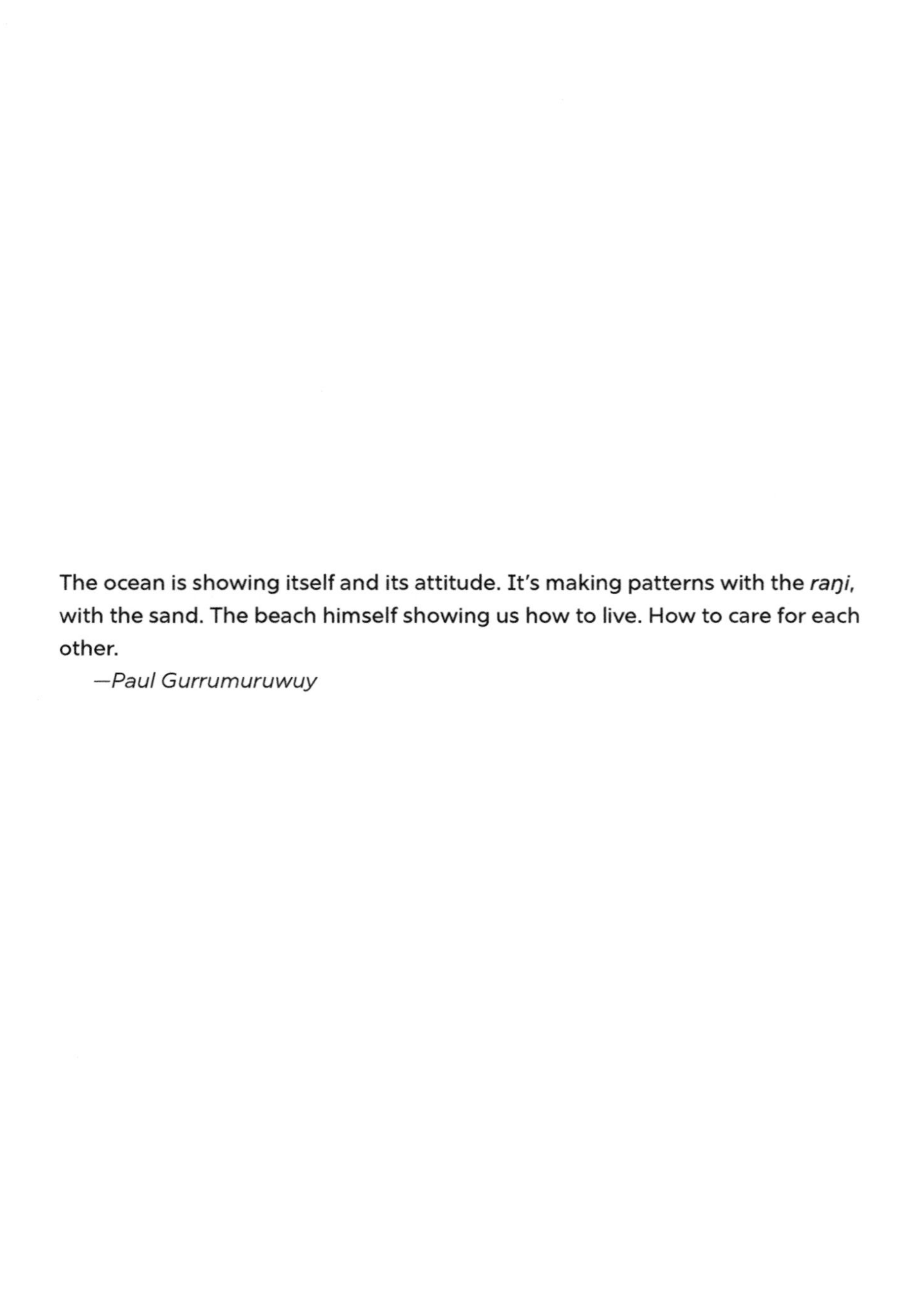

The ocean is showing itself and its attitude. It's making patterns with the *raŋi*, with the sand. The beach himself showing us how to live. How to care for each other.

—Paul Gurrumuruwuy

show yourself,
your values,
your way of living

so others can see

bala - räli djägamirr

you look after me, I'll look after you
we are holding each other

"I'm yours and you are mine. We belong to each other now"

the land
the colours
the patterns
the people

UNNUMBERED FIGURES 55–60 MIYARRKA MEDIA

Räl-manapanmirr means coming together. You can see that right there. That's what has to happen before *dhä-manapanmirr*. Before we can work together as one. You have to bring your different skills and talents, share them, join together as a team.

We are using the camera and computer, to show you something important. Because whether Yolŋu or balanda, it's the same as the *raŋi* [sand] and *gapu* [saltwater, sea]. We have to hold and care for each other. That's life.

—*Paul Gurrumuruwuy*

In times of profound ecological transformation and ongoing struggles for social and environmental justice, Miyarrka Media offers quiet lessons in relational aliveness. Rather than explaining this ethos to others, we use image and voice experiments to reach out directly from Country to connect with our audiences, and in turn, to stir their own capacity to care for their own places.[6]

Dhä-manapanmirr guides our commitment to work together across difference, providing an understanding that collaborative processes entail the reciprocation of respect, while presuming that under the right circumstances, and with the right team, knowledge can be gathered and assembled as a means of responding to shared concerns. Rather than translating or explaining one culture to another, we try out different ways of bringing together images, voices, ideas, perspectives, and feelings in ways that foster connections across separations of language, geography, history, and culture—while amplifying the foundational force of Country as a source of knowledge, care, and more-than-human mutuality.

While the emphasis in this project is on joy and renewal, long histories of colonial disregard and brutal impositions—including legal and bureaucratic processes that continue to overwrite Yolŋu law—provide an underlying charge of dissonance and exhausting dispute that quietly fuels our work. However, in keeping with a Yolŋu emphasis on the social significance of making relationships and avoiding direct critique, this is rarely made explicit.

Enlivened through our own ongoing experiences of *dhä-manapanmirr,* our team of Yolŋu and balanda work together, to model new ways to activate and share a lived ethos of reciprocity and responsibility for worlds falling apart *and* coming together. Taking our time, developing our own processes of curatorial care to the ways that the images, sounds, texts, and ideas can be made to visibly hum within digital screens, we attempt to bring messages from Country to strangers—who, we suspect, have their own deep connections to place. Or at least a yearning for this.

Gurrumuruwuy doesn't expect balanda to care particularly about his beach. He wants them to care more deeply, and more actively, for their own.

APPENDIX

Feral Atlas and *Field Guide*

Ideally, this book should be read together with the digital project, *Feral Atlas: The More-Than-Human Anthropocene* (www.feralatlas.org). While each can be appreciated on its own, the book and the atlas enrich each other. *Feral Atlas* is a collaborative project involving scientists, scholars, artists, poets, and vernacular and Indigenous observers of the Anthropocene. Its seventy-nine field reports offer a granular view of the Anthropocene; the atlas also offers an architecture for conceptualizing the Anthropocene in grounded, patch-based terms. The atlas presents the empirical materials and structural models that give life to the analysis in this book.

The *Field Guide* interprets the atlas, and vice versa. Each offers a series of sites, observations, thought experiments, and genre-stretching descriptive practices that insistently reach beyond the potentially homogenizing press of the planetary as a single scale for insight and action. Indeed, it's our hope that bringing *Feral Atlas* and *Field Guide* into dialogue will have its own generative effects. The following tips on using the website may be useful even for those who think they know the project back to front. There are hidden pleasures. And while we begin with basic user guidelines, the instructions spread into more mysterious corners.

Begin on the project's cover page. The cover is our team's first natural history illustration: Feifei Zhou has sketched a digital ecology of your computer screen in the time of COVID-19. Look for the small spiked spheres, which represent the viruses that might still be stuck on your screen. Around them are necrotic cells (not to scale). The feral world the atlas describes begins in front of you in the microbiome of your screen. Before you leave this page, you might scroll down to note the long and rich list of the atlas's contributors. When you are ready, click "Enter" to find yourself on the *Feral Atlas* landing page.

Digitally savvy users can click one of the floating feral entities (for example, Sudden Oak Death, Antibiotics, or Carbon Dioxide) to begin an interactive adventure. Other users should pause first, not only to admire the variety of beings represented in the atlas, but also to appreciate the tools available right from the landing page. Below the screen, you see three parallel lines; clicking those lines opens "the drawer," which includes instructions as well as links to readings. On the right side of the top of the inside of the drawer, you will see a link directly to the Reading Room, which includes framing essays for the whole digital project (Figure 61).

Return for a moment to the landing page for further orientation. On the top left of every screen, including the landing page, you should see a rotating brass key (Figure 62). This is a small joke: A "key" is the

FIGURE 61. **Screenshot of the drawer on the *Feral Atlas* landing page, with the Reading Room circled.**

information needed to read a map, but here we have created a door key. Clicking that key brings you to the Super Index, that is, the index of indices (Figure 63). The first screen of the Super Index is an interactive listing all the feral entities as well as the analytic categories through which the atlas helps you explore these entities. Roll your cursor over any entity, and the author or authors of reports on that entity will appear; click on an author's name and you will be taken directly to the field report on that entity. (Floating behind the listed field reports are categories that are not in *Feral Atlas*—but could be. We thus gesture beyond classification to possibility.) Clicking field reports from the index is one way to use the atlas, but it loses the experience of the framing categories; we suggest that you use this mode only after you have experienced the Anthropocene Detonators and Tippers that you will see if you click on an entity from the landing page.

If you scroll down in the Super Index, you will find the Reading Room, with links to all the framing essays (Figure 64). Several essays were commissioned from key contributors to Anthropocene discussions; take note of the section marked "Luminary Essays." These are written by eminent contributors to Anthropocene framings: geographers Simon Lewis and Mark Maslin; biologist David Richardson; writer Amitav Ghosh; historian Sven Beckert; climate scientist Will Steffen; and an-

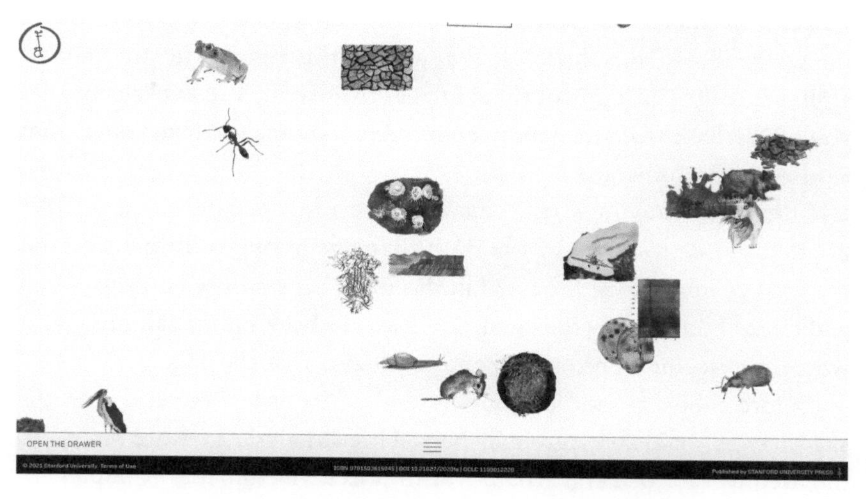

FIGURE 62. **Screenshot of the landing page with the key circled.**

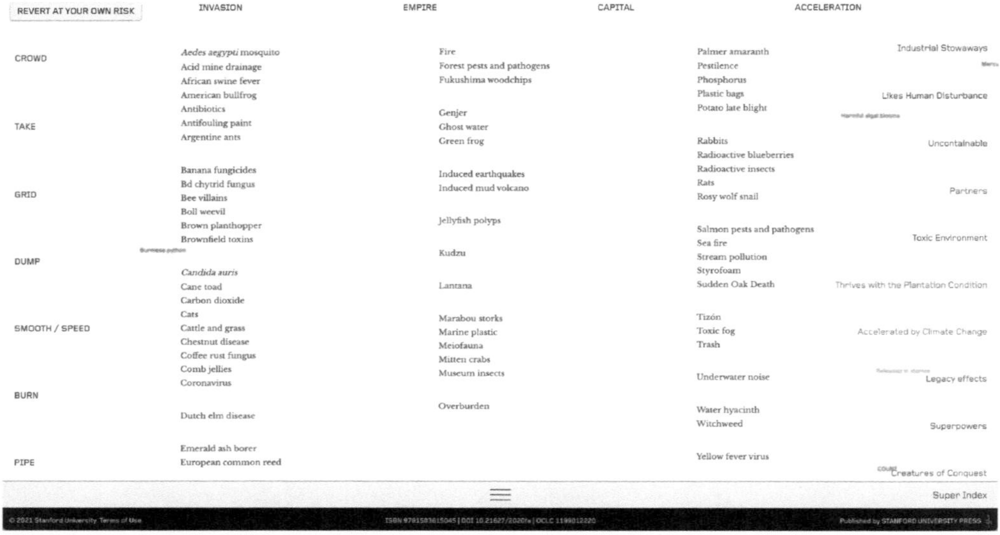

REVERT AT YOUR OWN RISK
INVASION
EMPIRE
CAPITAL
ACCELERATION

CROWD
TAKE
GRID
DUMP
SMOOTH / SPEED
BURN
PIPE

Aedes aegypti mosquito
Acid mine drainage
African swine fever
American bullfrog
Antibiotics
Antifouling paint
Argentine ants

Banana fungicides
Bd chytrid fungus
Bee villains
Boll weevil
Brown planthopper
Brownfield toxins

Candida auris
Cane toad
Carbon dioxide
Cats
Cattle and grass
Chestnut disease
Coffee rust fungus
Comb jellies
Coronavirus

Dutch elm disease

Emerald ash borer
European common reed

Fire
Forest pests and pathogens
Fukushima woodchips

Genjer
Ghost water
Green frog

Induced earthquakes
Induced mud volcano

Jellyfish polyps

Kudzu

Lantana

Marabou storks
Marine plastic
Meiofauna
Mitten crabs
Museum insects

Overburden

Palmer amaranth
Pestilence
Phosphorus
Plastic bags
Potato late blight

Rabbits
Radioactive blueberries
Radioactive insects
Rats
Rosy wolf snail

Salmon pests and pathogens
Sea fire
Stream pollution
Styrofoam
Sudden Oak Death

Tizón
Toxic fog
Trash

Underwater noise

Water hyacinth
Witchweed

Yellow fever virus

Industrial Stowaways
Likes Human Disturbance
Uncontainable
Partners
Toxic Environment
Thrives with the Plantation Condition
Accelerated by Climate Change
Legacy effects
Superpowers
Creatures of Conquest
Super Index

© 2021 Stanford University. Terms of Use
Published by STANFORD UNIVERSITY PRESS

FIGURE 63. **Screenshot of the top of the Super Index page.**

thropologist Karen Ho.[1] Also note that another of the Reading Room links, Contributors (under "Indices of Contributors and Contributions") brings you to the list of all the authors, and you can go directly to a contributor's work from that list. There are also discussions of teaching and sample syllabi using *Feral Atlas*.

Scroll farther toward the bottom of the Super Index, and you will see the watercolors from the landing page, but this time static and labeled alphabetically. Click any of these images to have the full experience that you would have from the landing page. After you have clicked once, that icon will turn full color.

"Revert at your own risk" means "return to the landing page." We use that phrase to remind users of the dangers of the Anthropocene and the feral collaborators discussed in the atlas. Let's proceed now as if you had clicked "revert at your own risk"; we are back on the landing page, where watercolor icons of feral entities float across the page.

Before choosing (and clicking) a feral entity, a few more tips on the site's navigational structure may help. Three analytic axes are built into the architecture of *Feral Atlas*: Anthropocene Detonators (Chapter 6), Tippers (see Chapter 4), and Feral Qualities (Chapter 10). Anthropo-

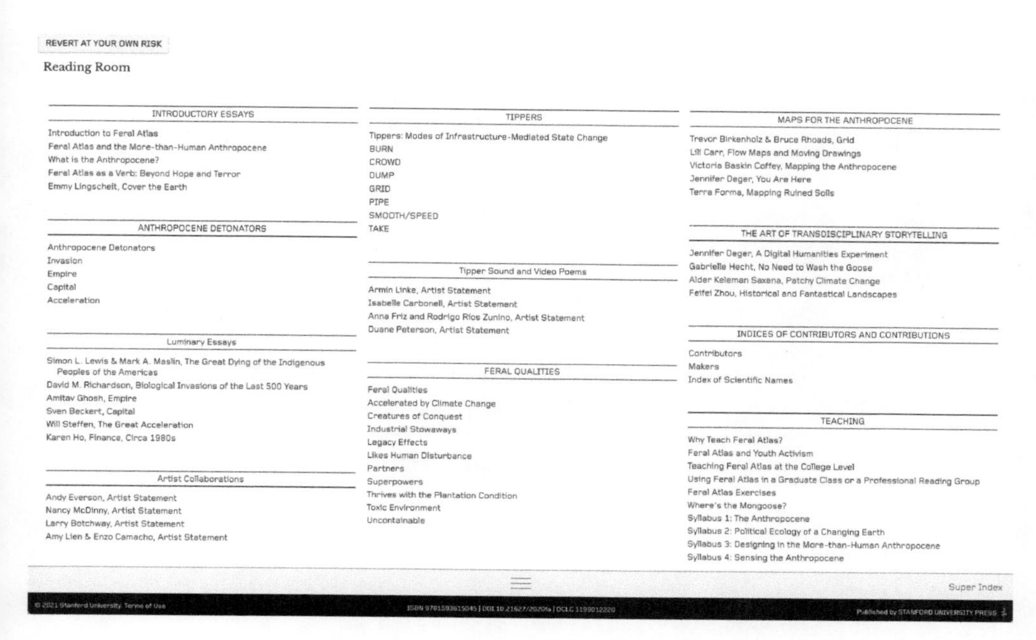

FIGURE 64. **Screenshot of the Reading Room page in the Super Index.**

cene Detonators are brought to life through historical collages, which we call Anthropocene Detonator landscapes (Chapter 11). Detonator landscapes show the historical conjunctures that give rise to powerful and dangerous infrastructure-building programs. Find these by clicking on a feral entity on the landing page. You'll see an explosion—the detonation—and then you will arrive at the Detonator landscape. Look for the red dot: That's the feral entity you chose (in Figure 65, the emerald ash borer). Zoom in and out as you would for Google Maps. Black dots are other feral entities you might want to follow. Roll over a dot to see the feral entity it hides. You can explore the Detonator landscape as a treasure hunt.

Clicking the feral entity on the Detonator landscape brings you to the Tippers page, which is explained in part through short videos, which we call video poems. The video poems show you the force of the kinds of work we expect from imperial and industrial infrastructures (Chapter 4). Figure 66 is a screenshot from a video poem for BURN, one of the Tippers. The infrastructures associated with these kinds of work have disrupted living ecologies. At the bottom of the screen, you will find a

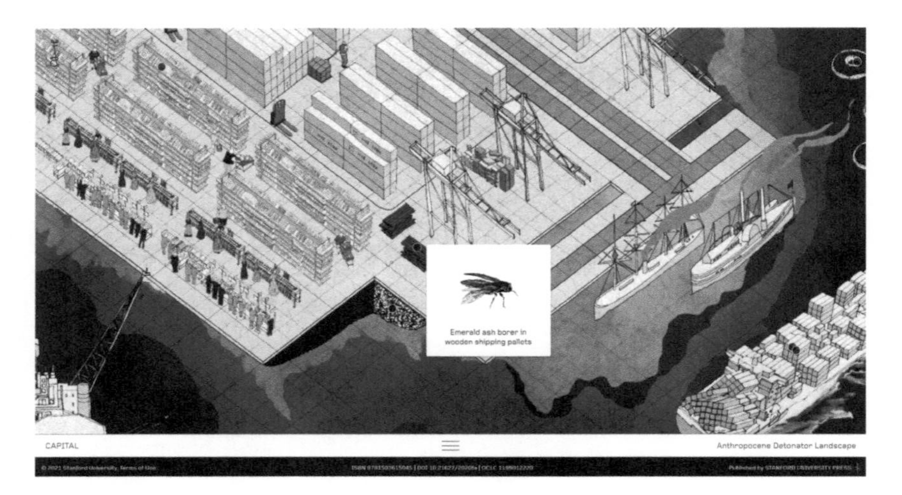

FIGURE 65. **Crop of *Capital* Detonator landscape page, with emerald ash borer highlighted.**

carousel of feral entities caught up in this Tipper. At the top right, you see a box that allows you to proceed to the field report about the feral entity you have chosen.

"Feral Qualities" are attunements made by feral entities to human infrastructures (Chapter 10). Feral Qualities are the hardest of the axes of analysis to find because they are located within the field reports, where they are indicated by interactive colored lines, found either on the left side of the screen, outside the text, or at the bottom of the main text, before the references (in Figure 67, on the left side of the report). Each color indicates one mode of attunement, one Feral Quality. For example, red is the color we have given to "Superpowers," that is, the rapid evolution or adaptation through which organisms change, becoming successful in the midst of human disturbance. If you click the Feral Qualities line, new text opens (Figure 68). The relevant section of the field report is highlighted. In the margin, a colored link appears that will bring the user to the framing text that explains that Feral Quality. The editors make additional comments— for example, describing just what Superpowers were gained or how these compare to effects described elsewhere in the atlas. A series of links is offered so users can connect to other field reports in which a feral entity develops Superpowers. The reader can move across the collection of field reports by connecting varied feral en-

FIGURE 66. **Screenshot of *First Oil Well*, Jabal al-Dukhan, Bahrain. 2014. BURN Tipper video poem.**
ARMIN LINKE

tities through their Feral Qualities. Feral Qualities are also a good place to explore for the poetry hidden in *Feral Atlas*.

Field reports on feral entities can be found in several ways:

1. Click an image on the landing page and follow it through Anthropocene Detonator landscapes and Tipper video poems to the field report.

2. Start with the interactive index in the Super Index (found by clicking the key); find the feral entity, hover, and click the author's name.

3. Find the Contributors' Index in the Super Index and click the contributor's name.

4. Scroll down to the bottom of the Super Index and click the appropriate image, which works just like the landing page except that the feral entities are quietly lined up in alphabetical order.

You can also enter the name of the entity and/or the author plus "Feral Atlas" into your browser—without opening the atlas at all. Each of these methods takes you to a field report. Meanwhile, analytic essays and analysis through art abound in the atlas.

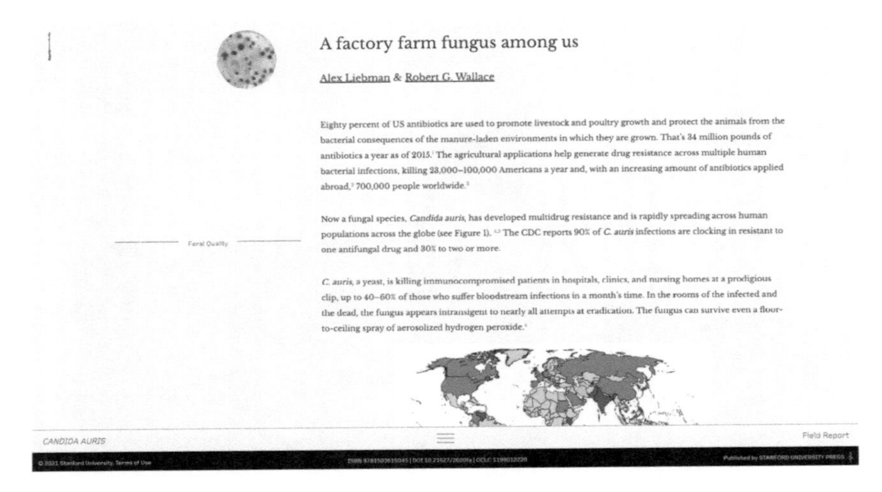

FIGURE 67. **A Superpower Feral Quality line accompanies Alex Liebman and Robert Wallace's report on *Candida auris*.**

FIGURE 68. **Clicking the Feral Qualities line brings up an explanation of the Feral Quality in relation to the field report.**

There is a great deal of material to read, look at, and listen to on the site, and a lot of opportunities for interactive play. Taken together, this book and the site guide a journey into the patchy Anthropocene.

ACKNOWLEDGMENTS

In this—as, perhaps, in most projects—the list of people who have made our work possible and our thinking more vibrant exceeds what is feasible to detail here. This list is just a beginning.

The digital project that served as the foundation for this book, *Feral Atlas: The More-Than-Human Anthropocene*, was a product of Aarhus University Research on the Anthropocene (AURA), itself supported by a five-year Niels Bohr Grant from the Danish National Research Foundation. We are indebted to the entirety of the AURA team as well as Aarhus University faculty and staff for hosting this endeavor. In particular, we thank Nils Bubandt, Elaine Gan, Mia Korsbaek, Peter Funch, Ton Otto, Jens Seeberg, Jens-Christian Svenning, and Heather Swanson. *Feral Atlas* also benefited from funding provided by the Aarhus University research subvention fund; and by Northern Arizona University. The project also received institutional and moral support from the Yale Program in Agrarian Studies and the School of Architecture at the Royal College of Art in London, and the Cairns Institute at James Cook University. Friederike Sundaram, our editor at Stanford University Press, provided valuable feedback on the digital project, and Jovan Maud offered wonderful skills as copy editor. Art Camp and Schoooool did the digital magic that made the project possible. A full list of project collaborators—including writers, artists, editors, and much more—can be found online at https://feralatlas.org/.

To all the contributors and facilitators of *Feral Atlas*, we extend our gratitude and appreciation. Two talented scholars and visual designers, Lili Carr and Victoria Baskin Coffey, made extensive contributions to that work, and our understandings continue to be indebted to their efforts. Videographers Armin Linke and Isabelle Carbonell were inspired—and patient—collaborators.

In moving from digital project to book, Jean Thompson Black, Paulla Ebron, Donna Haraway, Rusten Hogness, Inditian Latifa, Bruno Latour, Sumanth Prabhaker, Cosima Reichenbach, and Jane Sloan gave invaluable advice. Kathryn Chetkovich has our enduring gratitude for her assistance in sculpting the manuscript. The print book has received support and critical insight from our editor Kate Wahl.

In Chapter 5, the research referenced in Mexico was funded by an EPA Star Fellowship; and the research discussed in Bolivia was funded by an NSF Doctoral Dissertation Improvement Grant in Geography (#1131153). Research in both sites benefited from additional support from Yale University.

Chapter 7 originated as the Eric Wolf lecture of 2016; Anna Tsing is grateful to Andre Gingrich and his colleagues at the Austrian Academy of Sciences. Chapter 8 originated as the Marett Memorial Lecture at Exeter College of Oxford University in 2017; Tsing is grateful to David Gellner and his colleagues there. Tsing is also grateful to colleagues who read these papers for *Current Anthropology* and for the *Journal of the Royal Anthropological Institute*, respectively, with apologies that she stepped away from the revision process to attend to what seemed at the time to be more pressing obligations to *Feral Atlas*. Chapter 9 took varied oral forms but recently was a "Futures Lecture" at Aarhus University in 2022; for this invitation, Tsing is grateful to Nils Bubandt and his colleagues at Aarhus.

The research discussed in Chapter 12 was partially funded by the Australian Government through the Australian Research Council (project number SR200200346) in partnership with Goŋ-Däl Aboriginal Corporation and Gapuwiyak Culture and Arts.

Several colleagues and assistants lent logistic support to this project, including Kelly J. Baer, Aaron Glass, and Trevor van Weeren. Support for continuing research and writing have been provided by Northern Arizona University, the Northern Institute and the Center for Creative Futures at Charles Darwin University, Aarhus University, and the University of California, Santa Cruz.

We will miss—and remember—Lesley Stern, Bruno Latour, Will Steffen, Dieter Bruneel, and Temel Oğuz.

NOTES

Introduction

1. This example is drawn from Scott Gilbert's essay in *Feral Atlas: The More-Than-Human Anthropocene* (2021).

2. David Foster and colleagues' (2014) account of the fate of eastern hemlock offers a vivid example of ecosystem change due to insect damage to a single tree species.

3. Many commentators continue to see increasing human numbers as the problem without attention to the political ecologies that give some humans greater ecological footprints than others. Yet human numbers are themselves dependent on political economies of reproduction, which change radically in relation to state exactions and property regimes. Human numbers respond, too, to imperial and industrial infrastructures.

4. While some writers insist that the term "more-than-human" can only mean "superhuman," we disagree—and follow the usage that has become common practice in the environmental humanities and science studies. Throughout this book, we use the term "more-than-human" to refer to humans plus nonhumans. By "nonhuman," we mean both living creatures, such as the red turpentine beetle, and nonliving entities, such as petrochemicals, volatile dust particles, or phosphorus fertilizers, the characteristics of which nonetheless have the capacity to bring about change in the world.

5. Tsing et al., 2019.

6. Hausdoerffer et al., 2021.

7. In digital media, our team drew initial inspiration from Maya Lin's *What Is Missing?*, created in 2009. https://www.whatismissing.org/

8. For forays across the social and natural sciences, as well as often the arts, see for example, Bruno Latour and Peter Weibel, eds., *Critical Zones: The Science and Politics of Landing on Earth* (2020); Thom van Dooren, *The Wake of Crows: Living and Dying in Shared Worlds* (2019); Robin Wall Kimmerer, *Braiding Sweetgrass: Indigenous Wisdom, Scientific Knowledge, and the Teachings of Plants* (2014); Anna Lowenhaupt Tsing et al., eds., *Arts of Living on a Damaged Planet: Ghosts and Monsters of the Anthropocene* (2017); Haus der Kulturen der Welt and the Max Planck Institute for the History of Sci-

ence, *Mississippi An Anthropocene River 2018–19;* Heather Davis and Etienne Turpin, eds., *Art in the Anthropocene: Encounters Among Aesthetics, Politics, Environments and Epistemologies* (2015).

9. For a discussion of the revitalization of natural history and an analysis of obstacles within the natural sciences, see Joshua Tewksbury et al. (2014).

10. Some examples: for non-Western natural histories, Federico Marcon (2015); for Indigenous natural histories, Ian Saem Majnep and Ralph Bulmer (1977); for vernacular natural histories, Robert Macfarlane (2015).

11. For example, Bullard & Wright, 2012.

12. Caple, 2017.

13. Le Guin, 1989.

14. Gibson-Graham, 1996.

15. Bear et al., 2015.

16 Tsing et al., 2021.

17. For masculinity, see George Monbiot (2014). For chaos, see Richard Norton (2018) on feral cities.

18. For an important example of an interpretation of the Anthropocene as mastery, see John Asafu-Adjaye et al. (2015).

19. Nonhuman infrastructures are discussed in Chapter 8; these might be described as feral.

20. Wynter, 2015.

Part I

1. Between 2013 and 2018, Aarhus University Research on the Anthropocene (AURA), the research group that inspired *Field Guide*, explored this feral geography (Bubandt & Tsing, 2018).

2. Members of this substudy included Elaine Gan, Thiago Cardoso, Pierre du Plessis, Colin Hoag, and Nathalia Brichet. See Gan et al. (2018).

Chapter 1

1. Bullard & Wright, 2012.

2. Taylor, 2014.

3. In *Feral Atlas*, see the environmental justice analyses of Gabrielle Hecht (2021), on acid mine drainage, and John McNeill (2021) on the disease ecologies of Caribbean plantations. Each shows the importance of working across scales.

4. In *Feral Atlas*, artist Cornelia Hesse-Honegger (2021) documents the effects of radiation on insects, offering an important description of the Anthropocene.

5. Chakrabarty, 2009.

6. Chakrabarty, 2000.

7. Chakrabarty, 2021.

8. Ebron, 2021.

9. Beckert, 2015; Gilroy, 1995; Williams, 1944.

10. Paredes, 2021.

11. Wolford, 2021.

12. Ghosh, 2021b.

13 See Ghosh's essay "Empire" (2021a).

14. Wynter, 1971.

15. Wynter, 2015.

16. See Zalasiewicz et al., 2015; Steffen, 2021.

17. See also Tsing, 2019.

18. Carson, 1962.

19. Roy et al., 2021.

Chapter 2

1. Ellis et al., 2010.

2. For more on hyperobjects, see Timothy Morton (2013). For the critique, read Andreas Malm and Alf Hornborg (2014).

3. Quite a few reports in *Feral Atlas* explore routes through which feral beings travel. For example, Lionel Devlieger (2021) tracks mitten crabs between China and Europe, and Bettina Stoetzer (2021) tracks African swine fever as it has moved across Europe.

4. The first step in describing a patch is to establish the relations among the beings who happen to be there. This is a relational ontology, but one that does not require that only a functional network of human usefulness might count as "relations." From a description of the assemblage, a researcher can ask about landscape structure (or that of water or atmosphere), as this connects the patch to wider political ecologies. This is a moment to move beyond the patch to wider networks, human and nonhuman. But it is the patch that anchors field observation. It grounds our curiosity in a place, and (unlike networks defined around human usefulness) it allows us to stumble upon phenomena that we had no idea should hold our interest. If we want to study feral—that is, non-designed—effects of human infrastructures, we need to begin with the patch.

5. Fach et al., 2021.

6. See Frickel, 2021.

7. Sultana, 2022a.

8. For example, see Larkin, 2013; Anand et al., 2018.

9. Anand, 2018.

10. Vodopivec et al., 2021; see also Carbonell, 2021.

11. Funch, 2021.

12. For another everyday feral effect of urban infrastructure, see Frédéric Keck's (2021) report on the insects that eat museum holdings.

13. Gordon, 2021.

14. See Vann, 2021. Vann's history of Hanoi sewers is also told visually in the illustrated book, *The Great Hanoi Rat Hunt: Empire, Disease, and Modernity in French Colonial Vietnam* (Vann & Clarke, 2019).

15. Utopian claims have been built into cities by the kinds of infrastructures James C. Scott (1998) calls "high modernism."

16. For another *Feral Atlas* report on rats, as transported in ships, see Lesley Stern (2021).

17. See Davis et al., 2019. Note that non-American plantation legacies are equally important, including the nutmeg plantations that followed from the 1621 massacre in Banda, as explained in the discussion of Ghosh in Chapter 1.

18. Mintz, 1986.

19. Voeks & Rashford, 2013.

20. Mintz, 1960.

21. Li & Semedi, 2021.

22. Chao, 2022.

23. Li, 2017; Wolford, 2021.

24. Li & Semedi, 2021: vii.

25. Perfecto, 2021.

26 Perfecto, 2021.

Chapter 3

1. See Candace Fujikane's work with relational Kanaka and critical settler cartographies that sustain life: "Mapping abundance is a refusal to succumb to capital's logic that we have passed an apocalyptic threshold of no return" (Fujikane, 2021: 4).

2. For more on these themes, see David Turnbull and Helen Watson-Verran (1989) and Rasmus Winther (2020). See also geographers Rob Kitchin and Martin Dodge, who call for a mapping that shifts its foundational premise from "ontology (how things are) to ontogenesis (how things become)" (2007: 335).

3. For a discussion of the ways that mobility and animal studies track non-human entities in the constitution of spatial relations, see Timothy Hodgetts and Jamie Lorimer (2020).

4. Indigenous mapping is gaining increasing significance as a decolonizing research method. See, for example, Goeman Mishuana, *Mark My Words: Native Women Mapping Our Nations* (2013) and Reuben Rose-Redwood et al. (2020). See also Tony Syme's (2020) discussion of the need for critical consideration of the ways that GIS technologies potentially reproduce settler ontologies given the ontological assumptions built into the software itself. On decolonizing spatial practices more generally, see Eve Tuck and Marcia McKenzie, *Place in Research: Theory, Methodology, and Methods* (2015).

5. Jones & Jenkins, 2014: 17.

6. Christine Schranz argues that these technologies have imposed a new spatial order in which an egocentric view of the world dominates while "paradigmatically elevat[ing] the map to the interface between humans and computers (or systems) and space" (2021: 24).

7. In 1986 geographer J. Douglas Porteous warned of "the profound lack of curiosity" that remote sensing threatens to instantiate. "Do we need to be told that remote sensing is above all, remote?" he asked. "Could it be fieldwork demands skills (languages and empathy) which [researchers] are no longer willing to learn?" (1986: 251). Porteous advocated instead for forms of what he called "intimate sensing," a ground truthing method that entailed coming into close and unmediated relations with worlds "untidy rather than neat" (1986: 251).

For an exploration of intimate sensing of climate change and Indigenous knowledge practices, see Catherine Williams (2020). See also David Howes (2013) for a useful summary of "sensuous geography."

8. Lili Carr, map director for *Feral Atlas*, was a constant source of insight and inspiration for our maps. See Carr, 2021.

9 Snow & Witmer, 2021.

10 Seeberg, 2021.

11. Desimini, 2019: 25.

12 Brichet, 2021.

13 Weiss, 2021.

14. Massey, 2005: 4.

15 Cely-Santos, 2021.

16. Cely-Santos, 2021.

Part II

1. Swanson, 2015.

2. Zalasiewicz, 2017.

3. Bowker, 2000.

4. Hazen et al., 2012.

5. Chakrabarty, 2021.

6. Moore, 2016.

7. Uerta & Flores, 2022.

8. Friz & Ríos Zunino, 2021.

9. Wilkinson, 2005.

10. Rose, 2012.

Chapter 4

1. du Plessis, 2021.

2. du Plessis, 2018.

3. Livingston, 2019.

4. Vine, 2021.

5. For another report on California's "ghost water," see Drummond-Cole, 2021.

6. Blackbourne, 2007.

7. For California, see Nash, 2006.

8. Otto & Bubandt, 2011; Tsing et al., 2019.

9. Mauss, 1966.

10. Caple, 2021.

11. For another example of a body of water changed through eutrophication, see Agata Kowalewska (2021).

12. Williams, 2005.

13. McCully, 2001.

14. van Veen, 1962.

15. Appel, 2019.

16. Worster, 1992.

17. See Kate Brown (2013). Similarly, James Maguire (2021) discusses geothermal energy in Iceland, touted as a fully sustainable energy form. But, Maguire argues, when accelerated, it is a source of anthropogenic earthquakes.

18. Nikhil Anand et al. (2018) introduce the second of these definitions.

19. Gan, 2021.

20. Armin Linke, Isabelle Carbonell, Anna Friz, and Duane Peterson made most of the video poems in *Feral Atlas*. Their work inspired us.

21. Birkenholtz & Rhoads, 2021.

22. Milon & Zalasiewicz, 2021.

23 Milon & Zalasiewicz, 2021.

24. Ishida & Naito, 2021.

25. Nagy, 2021.

26. Reilly, 2009.

27. Steffen, 2021.

28. Brook et al., 2013.

Chapter 5

1. Pyne, 2021.

2. A note on how climate change features in *Feral Atlas* is warranted. In curating and editing the digital project, we aimed explicitly *not* to make climate our singular focus. Bracketing climate change in this way was useful analytically: It helped us to consider and articulate patterns in the many other forms of environmental disruption that make up the Anthropocene. When we did begin to work climate examples back into our analysis, this also helped us to see climate change not as a singular determinant force, but as an accelerator, potentiating and exacerbating feral processes already set in motion by human infrastructures. In the digital project, this is reflected in the group of field reports categorized under the Feral Quality "accelerated by climate change." Feral Qualities are discussed at greater length in Chapter 10 of this book.

3. Ingold, 2021.

4. See, for example, Nordgren Ballivián, 2011; Bolin, 2009; Magrath & Jennings, 2012.

5. For example, Myanna Lahsen has written extensively on the social construction of climate knowledge (see Lahsen, 2005, 2010, 2015); and Kenneth Broad and Ben Orlove (2007) have described the linkages between the local and the global drawn by emplaced actors experiencing climatic extremes. Farhana Sultana (2022b) offers an incisive outline of the argument that climate change develops in (and replicates) institutional contexts that rest on larger inequalities.

6. For more on critical climate justice, see Sultana (2022a).

7. Tsing et al., 2019.

8. This description draws on Karen Barad's theory of "intra-action" (2007), which is described at greater length in Chapter 9.

9. If you have an interest in social science theory, you may note internal contradictions in the term "material phenomenology." Our use of this term is outlined at greater length in Keleman Saxena (2021a). In brief, "materialism" is

often associated with Karl Marx—who was also a famous and vehement critic of phenomenology. However, our usage is more directly derived from recent writings on materiality, which urge a return to social science scholarship that considers the material characteristics of things to be part and parcel of (not ancillary to) their role in social phenomena (see also, Abrahamsson et al., 2015; Bennett, 2010; Keane, 2008; Miller, 2005). This in turn provides an underpinning for a first-person, situated, or phenomenological approach, which places lived experience at the center of analysis. While the "new" materialism can overlap with classical materialist critiques of capitalism, it also opens up other avenues of analysis. In this book, a "material phenomenology" aligns with these more recent theorizations of materiality—the kinds that urge us to sense the world from a situated perspective. This ideally begins by carefully noticing interactions of many kinds of human- and more-than-human agents (other species, climate and weather patterns, physical built infrastructures, etc.); but it notices these things while also attending to other aspects of the social (faith, politics, social media) and accounts for shifting relations among these diverse agents.

10. The idea of a "view from a patch" is a nod to ethnoecologist Virginia Nazarea, who famously wrote about ethnoecology as situated knowledge, that is a "view from a point." See Nazarea, 1999.

11. Keleman, 2005.

12. Keleman, 2010.

13. Blitzer, 2019.

14. Sultana (2015) uses the term "emotional political ecology" to describe an approach that considers how emotions and meanings matter for resource struggles.

15. Angé et al., 2018.

16. Keleman Saxena, 2021b; Keleman Saxena et al., 2016.

17. Fonseca & Sullivan, 2020; Goodluck, 2020.

18. Piltch et al., 2020.

19. O'Loughlin & Zaveri, 2020.

20. Tuttle et al., 2008.

21. Navajo Nation Government, 2023.

22. Ladyzhets et al., 2022.

23. Bennett-Begaye et al., 2021.

24. Fonseca, 2022a; Valencia, 2022.

25. Fonseca, 2022b.

26. Schwinning et al., 2008.

27. Brower, 2021.

28. Brower, 2021.

29. Dahm et al., 2015.

30. Allen, 2007.

31. Best Flagstaff Homes Realty, 2022.

32. "A Resolution of the City of Flagstaff," 2020.

33. See Oreskes & Conway, 2014.

34. Heglar, 2019.

35. See Hulme, 2011.

36. Dewan (2021) argues that an exclusive focus on climate as a driving force of social dislocation in the Bengal delta has led to a "misreading" of the importance of other societal issues, including gender norms and labor migration, to the detriment of development policy.

37. Tsing, 2005.

Chapter 6

1. An early proposal for the Anthropocene (Steffen et al., 2007) rested heavily on atmospheric carbon dioxide as a defining feature of epochal change. However, more recent formulations have considered a wider array of signatures of change, identifiable in geologic strata (Lewis & Maslin, 2015).

2. Crosby, 1972; Davis & Todd, 2017.

3. See Cronon, 1991; Fenn, 2002; McNeill, 2010.

4. Ghosh, 2021b; Haraway, 2016.

5. Beckert, 2015.

6. Moore, 2016.

7. McNeill & Engelke, 2014; Steffen, 2021.

8. Bessire, 2021.

9. For discussion of critical zones, see Bruno Latour and Peter Weibel (2020).

10. Cronon, 1991.

11. Tsing, 2015.

12. Crosby, 1972; Koch et al., 2019.

13. Whyte, 2017.

14. Anderson, 2004.

15. Ficek, 2021.

16. Keogh, 2011.

17. Lever, 2001.

18. Ashley, 2021.

19. Fenn, 2021; see also Fenn, 2014.

20. TallBear, 2016.

21. Whyte, 2017: 210.

22. Lewis & Maslin, 2015; Lewis & Maslin, 2021.

23. See, for example, de Sy et al., 2015.

24. Doody et al., 2021: 2.

25. For a general overview of zoogeochemistry, see Oswald Schmitz et al. (2018); for the role of invertebrates in nutrient cycling, see Matthew McCary and Schmitz (2021); for predator–prey interactions in relationship to nutrient cycling, see, among others, Alexander Flecker et al. (2019); Neil Hammerschlag et al. (2019); Shawn Leroux and Schmitz (2015); Alexandria Moore and Schmitz (2021); Schmitz et al. (2017); Christopher Wilmers and Schmitz (2016).

26. For other, scale-crossing, stories of human-led multispecies invasion, see Michael Hadfield (2021) on killer snails in Hawaii; Lucienne Strivay and Catherine Mougenot (2021) on rabbits in Australia; and Katy Overstreet (2021) on barn cats in the United States.

27. Zee, 2022.
28. Casid, 2004.
29. Bryant, 1995; Win & Kumazaki, 1998.
30. Geertz, 1963.
31. White, 1973.
32. Carney, 2001.
33. Biggs, 2011.
34. Stein, 2021; see also Stein & Luna, 2021.
35. Münster, 2021.
36. Devlieger, 2021.
37. Blackwell, 2021; see also Buttacavoli & McPherson, 2021.

Part III

1. Thompson, 2013.
2. Gilbert & Epel, 2015.
3. Marcia Bjornerud expresses the historical sentiment beautifully for geology: "Rocks are not nouns but verbs—visible evidence of processes: a volcanic eruption, the accretion of a coral reef, the growth of a mountain belt" (2018: 8).
4. Heiddeger, 1995.
5. See Plumwood, 1993; Haraway, 2008.

Chapter 7

1. This epigraph paraphrases the opening lines of Karl Marx's *The Eighteenth Brumaire of Louis Bonaparte* (1994, originally 1852), which chronicled the French revolutions of the eighteenth and nineteenth centuries. Marx argued that, while humans had the capacity to be historical actors, they were constrained by historically derived social structures, which conditioned their capacity to act. This section argues that, in considering nonhumans to be historical actors, one must also acknowledge their relationship to historically derived structuring forces, like capitalism, or climate.
2. Wolf, 1982.
3. This usage of "agility" borrows from, and stretches, that of Donna Haraway (2008), who uses the term in relation to the games she plays with her dogs.
4. Grossberg, 1986.
5. For the use of articulation theory to make North-South, cross-class, and cross-race connections, see Georges Dupré and Pierre-Philippe Rey (1973) and Stuart Hall (2018).
6. For another example, see Claude Meillassoux (1981).
7. For example, Hardt & Negri, 2003; Moore, 2015.
8. Holbraad & Pedersen, 2017.
9. For example, for salmon alone, see Karen Hébert (2015), Marianne Lien (2015), and Heather Swanson (2022).
10. Gopal, 1987.
11. Gopal, 1987: 54.
12. Buker, 1982: 421.
13. Pieterse, 1997.

14. However, see Trinidad Téllez et al. (2008) for the situation on the Iberian Peninsula.

15. Brij Gopal (1987: 98–99) and Téllez et al. (2008: 47) report figures of 400–700 tons per hectare per day of gained biomass.

16. Alimi & Akinyemiju, 1991.

17. Gopal, 1987: 167.

18. Kadono, 2004: 164.

19. Gopal, 1987: 50.

20. Gopal, 1987: 49.

21. Tinker, 1974: 747.

22. Zhang et al., 2010.

23. Wolf, 1982: 326–327.

24. Dupré & Rey, 1973; Wolf, 1982: 220–224.

25. Marx, 1992.

26. M-C-M: Money is used to buy commodities, which in turn are sold to gain money. This formulation is laid out in Marx's *Capital, Volume 1* (1992).

27. Freidberg, 2010.

28. The shifting mortgage and insurance markets for Japanese knotweed–infested homes in the UK are documented on many home insurance websites (for example, Bickers Insurance, 2015). Meanwhile, the internet is full of scary videos about Japanese knotweed; for an example see TCMltd (2010) on YouTube.

29. John P. Bailey and Ann P. Conolly (2000) offer an excellent overview of this history.

30. Bailey & Conolly, 2000: 100.

31. Casid, 2004.

32. Bailey & Conolly, 2000: 105.

33. For noxious chemicals, see Craig Murrell et al. (2011).

34. Barney et al., 2006.

35. John Bailey (2010, 2013) discusses the various hybrids, as well as the non-weedy nature of Japanese knotweed in its home location.

36. Bailey, 2013.

37. Barney et al., 2006.

38. The polyploid condition of the original female specimen is thought to have nurtured its many hybrid offspring by protecting them from recessive deficiencies; see Bailey, 2013.

39. Beckert, 2015; see also Beckert, 2021.

40. James Giesen (2011), the source for these statements, divides the South into different regional complexes, each of which showed different dynamics. My overview is not intended to diminish the importance of this heterogeneity.

41. Scholars of the U.S. South have argued over the impact of the boll weevil, with judgments ranging from negligible to extreme. Fabian Lange et al. (2009) contest the "negligible" argument with county-level data, which shows dramatic and persistent impacts for cotton producers.

42. Soluri, 2021; see also Chapter 9 of this book.

43. Ledbetter, 2021.

44. A "square" is a cotton bud.

45. Apocryphal stories of testing boll weevils circulated widely. Giesen (2011: 179, n.1) reports one farmer as saying:

> "Tain't no sense tryin' to fight that devil. I corked up a lot of 'em in pure alcohol and kept 'em for two hours. They come out staggering drunk, and with a mighty good appetite. Then I sealed 'em in a tin can an' throwed 'em in the fire. When the can melted, them red-hot bugs flew out and burnt my barn."

46. Since plants want to attract insects to their pollen, they do not put poisonous chemicals in pollen, and it has been hard for them to evolve a good defense against pollen-eating insects. See Jones, 2001.

47. Haney et al., 2009; Jones, 2001.

48. Stadler & Buteler, 2007: 211.

49. Dickens, 1986.

50. de Ribeiro et al., 2010.

51. Stadler & Buteler, 2007: 207.

52. Stadler & Buteler, 2007: 214.

53. Lacelli, 2004.

54. Hayes et al., 2008; Joshi & Sebastian, 2006.

55. Tsai et al., 2016.

56. Tsai, 2016.

57. Moore, 2015.

58. For the critique of anti-capitalist heroics, see Gibson-Graham (1996).

59. Gibson-Graham & Miller, 2015.

60. See Alfred Crosby (1972) on the transfer of organisms from Europe to the New World.

Chapter 8

1. The anthropological discussion of ontology looks beyond representations in knowledge production to the modes of being and world-making in which subjects and objects are embedded (see Kohn, 2015). In this chapter, I use the term "ontology" primarily because it is an *emic* term, that is, a term used by my scholarly interlocutors. Moving beyond a particular interlocutor, I combine the discussion of "historical ontologies" (Stoler, 2009) and of "multiple ontologies" (e.g., Mol, 2007). In this, however, I return to the articulation theories that draw together Part III of this book. It seems fair to note that, rather than an expansion of the use of ontology for anthropologists, this is an exploration of cautions and limits.

2. See Iqbal, 2010; Iqbal, 2021.

3. Marris, 2013.

4. Evans-Pritchard, 1950: 121.

5. Evans-Pritchard, 1950: 120.

6. Evans-Pritchard, 1950: 123.

7. See Barth, 2012. In the United States in 1950s, muscular claims of normative law-making appealed more than history. Talcott Parsons (1951) divided up turf across the disciplines, allowing social scientists to reimagine their material as data for synchronic hypothesis testing. George Murdock (1957) pioneered a big data approach to comparison.

8. Wolf, 1982.

9. Diouf, 2014.

10. Taussig, 1989.

11. Taussig, 1989: 20.

12. Mintz, 1986.

13. Viveiros de Castro, 2015.

14. Viveiros de Castro, 2014: 89.

15. Taussig, 1989.

16. de la Cadena, 2015.

17. Price, 2002.

18. Price, 1990.

19. Rosaldo, 1980.

20. Trouillot, 2011.

21. Stoler, 2009.

22. Stoler, 2009: 4.

23. Mol, 2007.

24. Williams, 2006.

25. Cosgrove, 1985.

26. Carse, 2019.

27. Carse, 2019: 208.

28. Carse, 2019: 214.

29. Mathews, 2018.

30. Evans-Pritchard's patterns, however, might not include expressions in landscape. In his 1950 lecture, he was particularly interested in patterns of intelligibility to human historical actors. To stretch this to include that intelligibility that arises out of building ways of life, human and nonhuman, is challenging but not impossible.

31. Morita, 2017.

32. Morita, 2017: 3, italics in original.

33. Fulmer, 2009.

34. Morita, 2016a.

35. Morita & Jensen, 2017.

36. Morita, 2016b.

37. Viveiros de Castro, 2004.

38. Mol, 2007.

39. Iqbal, 2010; Iqbal, 2021.

40. Iqbal, 2010: 128.

41. Sen, 1982.

42. E.g., Gopal, 1987.

43. Gopal, 1987: 261–262.

44. Gopal, 1987: 274.

45. Iqbal, 2010: 157.

46. Thomas Albright and co-authors (2004) used remote sensing to measure changes in water hyacinth infestation in two East African lakes. Varied strategies were attempted to thwart water hyacinth, including the introduction of specialized predator weevils. However, the sharpest declines seemed to have followed an El Niño year water influx that merely swept the plants away—just as rough waters did in the plant's Amazonian homeland.

47. Williams, 2006.

48. Barrett et al., 2008.

49. Zhang et al., 2010.

50. Barrett, 1977.

51. Barrett, 1979.

52. Wright & Jones, 2006.

53. Carse, 2019.

54. Price, 1990.

55. Geenen, 2012.

56. Adrian Williams (2006: 3) reports that two water hyacinth plants in the Democratic Republic of the Congo produced 1,200 daughter plants in four months.

57. Francaviglia, 1978.

58. Kohn, 2013.

59. Tsing, 1993.

60. Sahlins, 1985.

61. Spivak, 1988.

62. Strathern, 1987.

Chapter 9

1. It is hard not to think of Hayao Miyazaki's post-apocalyptic fantasy, *Nausicaä of the Valley of the Wind*, in which remaining humans eke out survival in the dust of toxic fungal blooms (Miyazaki, 1984).

2. Strathern, 2001.

3. Tsing, 2015.

4. Fisher et al., 2012.

5. Institute of Medicine, 2011; Tsing et al., 2017.

6. Gan & Tsing, 2018.

7. Quoted in Ramsbottom, 1937: 255.

8. Ramsbottom, 1937: 248–249.

9. Ramsbottom, 1937: 266.

10. Balasundaram et al., 2018.

11. Kauserud et al., 2007.

12. Bagchee, 1954; Singh et al., 1993.

13. Ramsbottom, 1937: 240.

14. Isaac, 1992: 44.

15. Other fungal diseases amplified by Anthropocene conditions: Chapter 2

introduced the coffee rust fungus (Perfecto, 2021); Chapter 5 mentioned potato early blight (Keleman Saxena, 2021b). Dutch elm disease (Brasier, 2021) is discussed in Chapter 10.

16. Garbelotto & Gonthier, 2013.

17. Doyle, 1985: 2.

18. Ullstrup, 1972.

19. Støvring Hovmøller, 2011; USDA, 2022.

20. Lieberei, 2007.

21. Grandin, 2010.

22. Garbelotto & Gonthier, 2013: 48.

23. Brasier, 2001; Brasier, 2021.

24. Stukenbrock, 2016: 104.

25. Depotter et al., 2016.

26. Benedict, 1981; Maloy, 1997.

27. Benedict, 1981: 19.

28. Benedict, 1981: 30.

29. Hoff et al., 2001.

30. Savory et al., 2015.

31. Brasier et al., 1999; Brasier & Jung, 2006.

32. See Garbelotto, 2021; see also Croucher et al., 2013.

33. Note that several other dangerous phytophthoras are featured in *Feral Atlas*. See Dieter Bruneel, Hanne Cottyn, and Esther Beeckaert (2021); Alder Keleman Saxena (2021b); Andrew Mathews (2021); and Bitty Roy and co-authors (2021).

34. See Kolby & Berger, 2021.

35. Farrer et al., 2011; Fisher et al., 2009.

36. E.g., Rosenblum et al., 2013.

37. O'Hanlon et al., 2018.

38. O'Hanlon et al., 2018: 623.

39. Whittaker & Vredenburg, 2011. Internal citations deleted.

40. Whittaker & Vredenburg, 2011. Internal citations deleted.

41. Fisher & Garner, 2007; Weldon et al., 2004.

42. See Snow & Witmer, 2021.

43. Kolby, 2014.

44. McCoy & Peralta, 2018.

45. Pegg et al., 2019.

46. Many of the issues raised in this section about fungi are also problems for bacteria. For soils full of antibiotics, see Lei Yang and Fangkai Zhao (2021). For the development of antibiotic resistance, see Jens Seeberg (2021).

47. Paredes, 2021.

48. Fisher et al., 2020.

49. Al-Hatmi et al., 2019.

50. Ma et al., 2010.

51. Batista et al., 2020.

52. Chowdhary et al., 2013.

53. Snelders et al., 2009.
54. Meis et al., 2016.
55. Burks et al., 2021.
56. Liebman & Wallace, 2021.
57. Chakrabarti & Sood, 2021: 6.
58. Chakrabarti & Sood, 2021: 4.
59. Liebman & Wallace, 2021.
60. Another interpretation blames global warming; as the world warms, the authors argue, fungi learn to adapt to the warmer settings of mammalian bodies. Nnaemeka Nnadi and Dee Carter write: "This yeast is considered the first 'novel' pathogen to have evolved in response to climate change" (2021: 2).
61. The significance of this transfer of nutrients between fields and forests is discussed further in Anna Tsing (2017).
62. Rackham, 2012.
63. Roy et al., 2021.
64. Tsing, 2015.
65. Barad, 2007.

Part IV

1. Miyarrka Media, 2019.

Chapter 10

1. In *Feral Atlas*, quite a few videographers and sound artists offered field reports, that is, descriptions of patches; see, for example, Isabelle Carbonell (2021); Anna Friz and Rodrigo Ríos Zunino (2021); Mike Sugarman (2021); and Chris Jordan (2021).
2. Mathews, 2021.
3. Alfred, 2021.
4. Contributing to the pile, *Feral Atlas* includes a second, quite different, report on salmon lice, by anthropologist Heather Swanson (2021).
5. Bogart & Putra, 2021.
6. Bubandt, 2021.
7. Bubandt, 2021. David Mackenzie (2021) offers a haunting video that captures the spirit of uncertainty.
8. Carroll, 1871.
9. Cohen et al., 2004; Holmes et al., 2006.
10. Adams & Sandbrook, 2013.
11. Law, 2015.
12. Mounzer, 2021.
13. Gilbert, 2021.
14. Liu, 2021.
15. See J. Brown, 2021; Fearnley & Lynteris, 2021; Hassell et al., 2021; Lin & Myers, 2021.
16. Wright, 2021.
17. Robin, 2019. To find this poem in *Feral Atlas*, scroll to the bottom of

Susan Wright's field report and click the bottom reddish line that says "Feral Quality."

18. Brasier, 2021.

19. Schmitz, 2021.

20. Fennelly, 2008. To find this poem in *Feral Atlas*, scroll to the bottom of Helene Schmitz's field report and click the bottom green line that says "Feral Quality."

21. Forseth & Innis, 2021.

22. For a quite different example of juxtaposition in *Feral Atlas*, consider its three reports on toxic fog: one concerning the cross-Pacific flow of pollution (Zee, 2021); one on a nineteenth-century killing fog in Belgium (Zimmer, 2021); and one on the relationship of pollution and lichens (Gabrys, 2021). Each adds to the others without being subsumed.

23. Skelly et al., 2021. See also Deborah Gordon's study of Argentine ants in California, which similarly argues against overbearing stereotypes about invasive ants (2021).

24. Doherty, 2021.

25. Spahr, 2011; Spahr, 2021.

26. Bassett & Spindel, 2021.

27. K. Brown, 2021.

28. It is this quality that has fueled criticisms that comparison is a colonial practice; see Ann Stoler (2001).

29. Law, 2019: 2.

Chapter 11

1 The concept of "unbuilding" has been used in a number of ongoing discussions in architecture, though not exactly in the ways that the term is deployed here. For example, Lionel Devlieger (2020) interprets "unbuilding" as demolition and deconstruction, and advocates for reusing building materials from unbuilding processes to create a more resource-efficient and less-extractive material economy and ecology. Others explore the idea of "not building" or "less building," such as in the research project "A Global Moratorium on New Construction" initiated by Charlotte Malterre-Barthes (2022), who argues for a radical reformation on the current capitalistic culture of the construction industry. As part of her research, she organized four online roundtable discussions, each exploring a different condition or approach towards a possible future of unbuilding: "Stop Construction?" hosted by the Harvard Graduate School of Design in April 2021; "Pivoting Practices" hosted by ETH Zurich in June 2021; "Non-Extractive Design" hosted by V–A–C Zattere in July 2021; and "Seeking Policy" hosted by Berlin Questions Conference, San Gimignano Lichtenberg, in August 2021.

2. Tsing, 2015.

3. Tsing, 2005.

4. Misrach & Orff, 2012.

5. McClintock, 2022.

6. See Aït-Touati et al., 2021.

7. *Ngurrara 1*; painting by Peter Skipper et al. (1996), held by the National Museum of Australia.

8. Lahoud, 2019.

9. "Ngurrara, The Great Sandy Desert Canvas," 2008

10. Batty, 1998.

11 Batty, 1998.

12. Lien & Camacho, 2021.

13. Botchway, 2021.

14. Coleman, 2021.

15. Le Guin, 1989.

16. Graber, 2022.

17. EPA, 2021.

18. Natural Resource Council of Maine, 2023.

19. American Rivers, 2023.

20. Rae, 2021.

21. hudsonriverkeeper, 2021.

22. Thomas-Blate, 2023.

23. Young, 2021.

24. Lacaton & Vassal, 1996.

25. Lacaton & Vassal, 1996.

26. Wainwright, 2021.

27. See Vodopivec et al., 2021.

28. See Roy et al., 2021.

29. Connuck, 2022.

30. Taijian, 2015.

31. Ren, 2021.

Chapter 12

1. Our work over the past fifteen years has been co-conceived and funded as research, rather than art or documentary. Gurrumuruwuy and I co-founded the Centre for Creative Futures at Charles Darwin University in 2022. I am an anthropologist and media maker by training. Gurrumuruwuy has worked nationally and internationally as a creative cultural mediator, actor, dancer, and performer.

2 Miyarrka Media, 2019: 42.

3 See also Armstrong et al., 2022.

4. Christie, 2009.

5. Miyarrka Media, 2019.

6. See also Gurrumuruwuy et al., in press.

Appendix

1. See Lewis & Maslin, 2021; Richardson, 2021; Ghosh, 2021; Beckert, 2021; Steffen, 2021; and Ho, 2021. Other important essays are also in the Reading Room, including work by Hecht, 2021; Aït-Touati et al. (the *Terra Forma* group), 2021; and members of the *Feral Atlas* team.

BIBLIOGRAPHY

Abrahamsson, S., Bertoni, F., Mol, A., & Martín, R. I. (2015). "Living with Omega-3: New Materialism and Enduring Concerns." *Environment and Planning D: Society and Space*, *33*(1), 4–19. https://doi.org/10.1068/d14086p

Adams, W., & Sandbrook, C. (2013). "Conservation, Evidence, and Policy." *Oryx*, *47*(3), 329–335.

Aït-Touati, F., Arènes, A., & Grégoire, A. (2021). "Terra Forma, Mapping Ruined Soils." In *Feral Atlas: The More-Than-Human Anthropocene*, edited by A. L. Tsing, J. Deger, A. K. Saxena, & F. Zhou. Stanford University Press. https://feralatlas.supdigital.org/index?text=terra-forma-mapping-ruined-soils&ttype=essay&cd=true

Albright, T., Moorhouse, T. G., & McNabb, T. J. (2004). "The Rise and Fall of Water Hyacinth in Lake Victoria and the Kagera River Basin, 1989–2001." *Journal of Aquatic Plant Management*, 42, 73–84.

Alfred, E. (2021). "Long Ago, My Ancestors Could Walk Across the River on the Backs of the Salmon." In *Feral Atlas: The More-Than-Human Anthropocene*, edited by A. L. Tsing, J. Deger, A. K. Saxena, & F. Zhou. Stanford University Press. https://feralatlas.supdigital.org/poster/my-grandparents-used-to-tell-stories-where-you-could-walk-across-the-river

Al-Hatmi, A., de Hoog, G. S., & Meis, J. F. (2019). "Multiresistant *Fusarium* Pathogens on Plants and Humans: Solutions in (from) the Antifungal Pipeline?" *Infection and Drug Resistance*, *12*, 3727–3737. https://doi.org/10.2147/IDR.S180912

Alimi, T., & Akinyemiju, O. (1991). "Effects of Water Hyacinth on Water Transportation in Nigeria." *Journal of Aquatic Plant Management*, *29*, 109–112.

Allen, C. D. (2007). "Interactions Across Spatial Scales Among Forest Dieback, Fire, and Erosion in Northern New Mexico Landscapes." *Ecosystems*, *10*(5), 797–808. https://doi.org/10.1007/s10021-007-9057-4

American Rivers. (2023). *Penobscot River: Ancestral Homeland in Recovery*. American Rivers. https://www.americanrivers.org/river/penobscot-river/

Anand, N. (2018). "A Public Matter: Water, Hydraulics, Biopolitics." In *The Promise of Infrastructure,* edited by N. Anand, A. Gupta, & H. Appel (pp. 155–172). Duke University Press.

Anand, N., Gupta, A., & Appel, H., eds. (2018). *The Promise of Infrastructure.* Duke University Press.

Anderson, V. (2004). *Creatures of Empire: How Domestic Animals Transformed Early America.* Oxford University Press.

Angé, O., Chipa, A., Condori, P., Ccoyo, A. C., Mamani, L., Pacco, R., Quispe, N., Quispe, W., & Sutta, M. (2018). "Interspecies Respect and Potato Conservation in the Peruvian Cradle of Domestication." *Conservation and Society, 16*(1), 30–40.

Appel, H. (2019). *The Licit Life of Capitalism: US Oil in Equatorial Guinea.* Duke University Press.

Armstrong, E., Gapany, D., Maypilama, L., Bukulatjpi, Y., Fasoli, L., Ireland, S., & Lowell, A. (2022). "Räl-Manapanmirr Ga Dhä-Manapanmirr—Collaborating and Connecting: Creating an Educational Process and Multimedia Resources to Facilitate Intercultural Communication." *International Journal of Speech-Language Pathology,* 24(5), 533–546.

Asafu-Adjaye, J., Blomquist, L., Brand, S., Brook, B., Defries, R., Ellis, E., Foreman, C., Keith, D., Lewis, M., Lynus, M., Nordhaus, T., Pielke, R., Pritzker, R., Ronald, P., Roy, J., Sagoff, M., Shellenberger, M., Stone, R., & Teague, P. (2015). *An Ecomodernist Manifesto.* Breakthrough Institute.

Ashley, R. N. (2021). "Before, Goannas Were Here Forever." In *Feral Atlas: The More-Than-Human Anthropocene*, edited by A. L. Tsing, J. Deger, A. K. Saxena, & F. Zhou. Stanford University Press. https://feralatlas.supdigital.org/poster/before-goannas-were-here-forever

Asiegbu, F. O., Adomas, A., & Stenlid, J. (2005). "Conifer Root and Butt Rot Caused by *Heterobasidion annosum* (Fr.) Bref. *s.l.*" *Molecular Plant Pathology, 6*(4), 395–409. https://doi.org/10.1111/j.1364-3703.2005.00295.x

Bagchee, K. (1954). "*Merulius lacrymans* (Wulf.) Fr. in India." *Sydowia, 8*, 80–85.

Bailey, J. (2010). "Opening Pandora's Seed Packet." *The Horticulturalist* (April), 21–24.

———. (2013). "Case Studies of Classic Examples of Hybridisation and Polyploidy: The Japanese Knotweed Invasion Viewed as a Vast Unintentional Hybridisation Experiment." *Heredity, 110*, 105–110.

Bailey, J. P., & Conolly, A. P. (2000). "Prize-Winners to Pariahs: A History of Japanese Knotweed *s. l.* (Polygonaceae) in the British Isles." *Watsonia, 23*(1), 93–110.

Balasundaram, S. V., Hess, J., Durling, M. B., Moody, S. C., Thorbek, L., Progida, C., LaButti, K., Aerts, A., Barry, K., Grigoriev, I. V., Boddy, L., Högberg, N., Kauserud, H., Eastwood, D. C., & Skrede, I. (2018). "The Fungus That Came in from the Cold: Dry Rot's Pre-Adapted Ability to Invade Buildings." *ISME Journal, 12*(3), 791–801. https://doi.org/10.1038/s41396-017-0006-8

Barad, K. (2007). *Meeting the Universe Halfway: Quantum Physics and the Entanglement of Matter and Meaning.* Duke University Press.

Barney, J. N., Tharayil, N., DiTommaso, A., & Bhowmik, P. C. (2006). "The

Biology of Invasive Alien Plants in Canada. 5. *Polygonum cuspidatum* Sieb. & Zucc. [= *Fallopia japonica* (Houtt.) Ronse Decr.]." *Canadian Journal of Plant Science*, *86*(3), 887–906. https://doi.org/10.4141/P05-170

Barrett, S. C. H. (1977). "Tristyly in *Eichhornia crassipes* (Mart.) Solms (Water Hyacinth)." *Biotropica*, *9*(4), 230–238. https://doi.org/10.2307/2388140

———. (1979). "The Evolutionary Breakdown of Tristyly in *Eichhornia crassipes* (Mart.) Solms (Water Hyacinth)." *Evolution*, *33*(1), 499–510. https://doi.org/10.2307/2407638

Barrett, S. C. H., Colautti, R. I., & Eckert, C. G. (2008). "Plant Reproductive Systems and Evolution During Biological Invasion." *Molecular Ecology*, *17*(1), 373–383. https://doi.org/10.1111/j.1365-294X.2007.03503.x

Barth, F. (2012). "Britain and the Commonwealth." In *One Discipline, Four Ways: British, German, French, and American Anthropology*, edited by F. Barth, A. Gingrich, R. Parkin, & S. Silverman (pp. 1–57). University of Chicago Press.

Bassett, T., & Spindel, C. (2021). "This Stowaway Plant Is Here to Stay." In *Feral Atlas: The More-Than-Human Anthropocene*, edited by A. L. Tsing, J. Deger, A. K. Saxena, & F. Zhou. Stanford University Press. https://feralatlas.supdigital.org/poster/this-stowaway-plant-is-here-to-stay

Batista, B. G., Chaves, M. A. de, Reginatto, P., Saraiva, O. J., & Fuentefria, A. M. (2020). "Human Fusariosis: An Emerging Infection That Is Difficult to Treat." *Revista Da Sociedade Brasileira de Medicina Tropical*, *53*, e20200013. https://doi.org/10.1590/0037-8682-0013-2020

Batty, D., dir. (1998). *Jila—Painted Waters of the Sandy Desert*. Rebel Films. https://rebelfilms.com.au/product/jila/

Bear, L., Ho, K., Tsing, A. L., & Yanagisako, S. (2015, March 30). *Gens: A Feminist Manifesto for the Study of Capitalism*. Society for Cultural Anthropology. https://culanth.org/fieldsights/gens-a-feminist-manifesto-for-the-study-of-capitalism

Beckert, S. (2015). *Empire of Cotton: A Global History*. Vintage Books.

———. (2021). "Capital." In *Feral Atlas: The More-Than-Human Anthropocene*, edited by A. L. Tsing, J. Deger, A. K. Saxena, & F. Zhou. Stanford University Press. https://feralatlas.supdigital.org/index?text=beckert-capital&ttype=essay&cd=true

Benedict, W. V. (1981). *History of White Pine Blister Rust Control: A Personal Account*. USDA Forest Service.

Bennett, J. (2010). *Vibrant Matter: A Political Ecology of Things*. Duke University Press.

Bennett-Begaye, J., Clahchischiligi, S., & Trudeau, C. (2021, June 8). "A Broken System: The Number of Indigenous People Who Died from Coronavirus May Never Be Known." *High Country News*. https://www.hcn.org/articles/indigenous-affairs-covid19-a-broken-system-the-number-of-indigenous-people-who-died-from-coronavirus-may-never-be-known

Bessire, L. (2021). *Running Out: In Search of Water on the High Plains*. Princeton University Press.

Best Flagstaff Homes Realty. (2022, April 20). *Flagstaff Real Estate Annual Market Trends*. https://bestflagstaffhomes.com/annual-trends/

Bickers Insurance. (2015, January 9). *Japanese Knotweed and Buildings, Mortgages and Insurance*. https://www.bickersinsurance.co.uk/about-us/latest-news/property-owners-news/japanese-knotweed-and-buildings-mortgages-and-insurance/

Biggs, D. (2011). *Quagmire: Nation-Building and Nature in the Mekong Delta*. University of Washington Press.

Birkenholtz, T., & Rhoads, B. (2021). "Grid." In *Feral Atlas: The More-Than-Human Anthropocene*, edited by A. L. Tsing, J. Deger, A. K. Saxena, & F. Zhou. Stanford University Press. https://feralatlas.supdigital.org/index?text=trevor-birkenholz-and-bruce-rhoads-grid&ttype=essay&cd=true

Bjornerud, M. (2018). *Timefulness: How Thinking Like a Geologist Can Help Save the World*. Princeton University Press.

Blackbourne, D. (2007). *The Conquest of Nature: Water, Landscape, and the Making of Modern Germany*. W. W. Norton.

Blackwell, S. B. (2021). "There Is No Escaping Underwater Noise." In *Feral Atlas: The More-Than-Human Anthropocene*, edited by A. L. Tsing, J. Deger, A. K. Saxena, & F. Zhou. Stanford University Press. https://feralatlas.supdigital.org/poster/there-is-no-escaping-underwater-noise

Blitzer, J. (2019, April 3). "How Climate Change Is Fuelling the U.S. Border Crisis." *The New Yorker*. https://www.newyorker.com/news/dispatch/how-climate-change-is-fuelling-the-us-border-crisis

Bogart, G., & Putra, G. (2021). "Weedy Invader Genjer Has a Tangled Political Past in Indonesia." In *Feral Atlas: The More-Than-Human Anthropocene*, edited by A. L. Tsing, J. Deger, A. K. Saxena, & F. Zhou. Stanford University Press. https://feralatlas.supdigital.org/poster/weedy-invader-genjer-has-a-tangled-political-past-in-indonesia

Bolin, I. (2009). "The Glaciers of the Andes Are Melting: Indigenous and Anthropological Knowledge Merge in Restoring Water Resources." In *Anthropology and Climate Change: From Encounters to Actions*, edited by S. A. Crate & M. Nuttall. Routledge.

Botchway, L. (2021). "Artist Statement." In *Feral Atlas: The More-Than-Human Anthropocene*, edited by A. L. Tsing, J. Deger, A. K. Saxena, & F. Zhou. Stanford University Press. https://feralatlas.supdigital.org/index?text=larry-botchway-artist-statement&ttype=essay&cd=true

Bowker, G. (2000). "Biodiversity Datadiversity." *Social Studies of Science*, *30*(5), 643–683, Article 5. https://doi.org/10.1177/030631200030005001

Brasier, C. M. (2001). "Rapid Evolution of Introduced Plant Pathogens via Interspecific Hybridization." *BioScience*, *51*(2), 123–133. https://doi.org/10.1641/0006-3568(2001)051[0123:REOIPP]2.0.CO;2

———. (2021). "Introduced Pathogens Can Evolve Rapidly, Increasing Their Virulence." In *Feral Atlas: The More-Than-Human Anthropocene*, edited by A. L. Tsing, J. Deger, A. K. Saxena, & F. Zhou. Stanford University Press. https://feralatlas.supdigital.org/poster/introduced-pathogens-can-evolve-rapidly-increasing-their-virulence

Brasier, C. M., Cooke, D. E. L., & Duncan, J. M. (1999). "Origin of a New *Phytophthora* Pathogen Through Interspecific Hybridization." *PNAS, 96*(10), 5878–5883. https://doi.org/10.1073/pnas.96.10.5878

Brasier, C. M., & Jung, T. M. (2006). "Recent Developments in Phytophthora Diseases of Trees and Natural Ecosystems in Europe." In *Progress in Research on Phytophthora Diseases of Forest Trees*, edited by C. M. Brasier, T. M. Jung, & W. Oßwald (pp. 5–16). Forest Research.

Brichet, Nathalia. (2021). "Cruise Ships Deliver Chemical Cocktails to Caribbean Marine Life." In *Feral Atlas: The More-than-Human Anthropocene*, edited by A. L. Tsing, J. Deger, A. K. Saxena, & F. Zhou. Stanford University Press. https://feralatlas.supdigital.org/poster/cruise-ships-deliver-chemical-cocktails-to-caribbean-marine-life

Broad, K., & Orlove, B. (2007). "Channeling Globality: The 1997–98 El Niño Climate Event in Peru." *American Ethnologist, 34*(2), 285–302. https://doi.org/10.1525/ae.2007.34.2.285

Brook, B. W., Ellis, E. C., Perring, M. P., Mackay, A. W., & Blomqvist, L. (2013). "Does the Terrestrial Biosphere Have Planetary Tipping Points?" *Trends in Ecology & Evolution, 28*(7), 396–401. https://doi.org/10.1016/j.tree.2013.01.016

Brower, M. (2021, September 12). "Flagstaff Addresses Flood Damage After Record Rains on Museum Fire Burn Scar." *AZ Central*. https://www.azcentral.com/story/news/local/arizona-weather/2021/09/12/flagstaff-addresses-flood-damage-after-record-rains-museum-fire-burn-scar/5622981001/

Brown, J. (2021). "Bullet Points." In *Feral Atlas: The More-Than-Human Anthropocene*, edited by A. L. Tsing, J. Deger, A. K. Saxena, & F. Zhou. Stanford University Press. https://feralatlas.supdigital.org/poster/coronavirus-stories-are-still-emerging

Brown, K. (2013). *Plutopia: Nuclear Families, Atomic Cities, and the Great Soviet and American Plutonium Disasters*. Oxford University Press.

———. (2021). "Chernobyl Is on Your Breakfast Table." In *Feral Atlas: The More-Than-Human Anthropocene*, edited by A. L. Tsing, J. Deger, A. K. Saxena, & F. Zhou. Stanford University Press. https://feralatlas.supdigital.org/poster/chernobyl-is-going-global

Bruneel, D., Cottyn, H., & Beeckaert, E. (2021). "Potato Late Blight Follows Crowding and Impoverishment." In *Feral Atlas: The More-Than-Human Anthropocene*, edited by A. L. Tsing, J. Deger, A. K. Saxena, & F. Zhou. Stanford University Press. https://feralatlas.supdigital.org/poster/potato-late-blight-follows-crowding-and-impoverishment

Bryant, R. L. (1995). "Shifting Under a Colonial Past." *Down to Earth*. https://www.downtoearth.org.in/coverage/shifting-under-a-colonial-past-27568

Bubandt, N. (2021). "Mud Overflows Boreholes, Politics, and Reason." In *Feral Atlas: The More-Than-Human Anthropocene*, edited by A. L. Tsing, J. Deger, A. K. Saxena, & F. Zhou. Stanford University Press. https://feralatlas.supdigital.org/poster/in-a-landscape-disturbed-by-mining-mud-overflows-boreholes-politics-and

Bubandt, N., & Tsing, A. (2018). "An Ethnoecology for the Anthropocene: How a Former Brown-Coal Mine in Denmark Shows Us the Feral Dynamics of Post-Industrial Ruin." *Journal of Ethnobiology, 38*(1), Supplement. https://doi.org/10.2993/0278-0771-38.1.001

Buker, G. (1982). "Engineers vs Florida's Green Menace." *Florida Historical Quarterly, 60*(4), 413–427.

Bullard, R. D., & Wright, B. (2012). *The Wrong Complexion for Protection: How the Government Response to Disaster Endangers African American Communities.* NYU Press.

Burks, C., Darby, A., Gómez Londoño, L., Momany, M., & Brewer, M. T. (2021). "Azole-Resistant *Aspergillus fumigatus* in the Environment: Identifying Key Reservoirs and Hotspots of Antifungal Resistance." *PLOS Pathogens, 17*(7), e1009711. https://doi.org/10.1371/journal.ppat.1009711

Buttacavoli, M., & McPherson, G. (2021). "Cargo Ship Noise Jeopardizes Ocean Life." In *Feral Atlas: The More-Than-Human Anthropocene*, edited by A. L. Tsing, J. Deger, A. K. Saxena, & F. Zhou. Stanford University Press. https://feralatlas.supdigital.org/poster/cargo-ship-noise-jeopardizes-ocean-life

Caple, Z. (2017). *Holocene in Fragments: A Critical Landscape Ecology of Phosphorus in Florida.* PhD dissertation. Department of Anthropology, University of California, Santa Cruz.

———. (2021). "Chemical Fertilizers Turn a Life-Bearing Element into an Ecological Menace." In *Feral Atlas: The More-Than-Human Anthropocene*, edited by A. L. Tsing, J. Deger, A. K. Saxena, & F. Zhou. Stanford University Press. https://feralatlas.supdigital.org/poster/chemical-fertilizers-turn-a-life-bearing-element-into-an-ecological-menace

Carbonell, I. (2021). "Polyps Are a Pluriverse." In *Feral Atlas: The More-Than-Human Anthropocene*, edited by A. L. Tsing, J. Deger, A. K. Saxena, & F. Zhou. Stanford University Press. https://feralatlas.supdigital.org/poster/polyps-are-a-pluriverse

Carney, J. A. (2001). *Black Rice: The African Origins of Rice Cultivation in the Americas.* Harvard University Press.

Carr, L. (2021). "Flow Maps and Moving Drawings." In *Feral Atlas: The More-Than-Human Anthropocene*, edited by A. L. Tsing, J. Deger, A. K. Saxena, & F. Zhou. Stanford University Press. https://feralatlas.supdigital.org/index?text=lili-carr-flow-maps-and-moving-drawings&ttype=essay&cd=true

Carroll, L. (1871). *Through the Looking-Glass and What Alice Found There.* Macmillan.

Carse, A. (2019). *Beyond the Big Ditch: Politics, Ecology, and Infrastructure at the Panama Canal.* MIT Press.

Carson, R. (1962). *Silent Spring.* Houghton Mifflin.

Casid, J. H. (2004). *Sowing Empire: Landscape and Colonization.* University of Minnesota Press.

Cely-Santos, M. (2021). "Only a Few Bee Species Thrive in the Killing Fields of Industrial Agriculture." In *Feral Atlas: The More-Than-Human Anthropo-*

cene, edited by A. L. Tsing, J. Deger, A. K. Saxena, & F. Zhou. Stanford University Press. https://feralatlas.supdigital.org/poster/only-a-few-bee-species-thrive-in-the-killing-fields-of-industrial-agriculture

Centers for Disease Control and Prevention (CDC). (2020, May 29). *Tracking Candida auris*. Centers for Disease Control and Prevention. https://www.cdc.gov/fungal/candida-auris/tracking-c-auris.html

Chakrabarti, A., & Sood, P. (2021). "On the Emergence, Spread and Resistance of *Candida auris*: Host, Pathogen and Environmental Tipping Points." *Journal of Medical Microbiology*, *70*(3). https://doi.org/10.1099/jmm.0.001318

Chakrabarty, D. (2000). *Provincializing Europe: Postcolonial Thought and Historical Difference*. Princeton University Press.

———. (2009). "The Climate of History: Four Theses." *Critical Inquiry*, *35*(2), 197–222.

———. (2021). *The Climate of History in a Planetary Age*. University of Chicago Press.

Chao, S. (2022). *In the Shadow of the Palms: More-Than-Human Becomings in West Papua*. Duke University Press.

Chowdhary, A., Kathuria, S., Xu, J., & Meis, J. F. (2013). "Emergence of Azole-Resistant *Aspergillus fumigatus* Strains Due to Agricultural Azole Use Creates an Increasing Threat to Human Health." *PLOS Pathogens*, *9*(10), e1003633. https://doi.org/10.1371/journal.ppat.1003633

Christie, M. (2009). "Engaging with Australian Indigenous Knowledge Systems: Charles Darwin University and the Yolŋu of Northeast Arnhemland." *Learning Communities: International Journal of Learning in Social Contexts*, 23–25.

Cohen, A., Stavri, P. Z., & Hersh, W. R. (2004). "A Categorization and Analysis of the Criticisms of Evidence-Based Medicine." *International Journal of Medical Informatics*, *73*, 35–43.

Coleman, C. (2021, September 23). "All Aboriginal Art Is Political: You Just Need to Learn How to Read It." *The Guardian*. https://www.theguardian.com/commentisfree/2021/sep/23/all-aboriginal-art-is-political-you-just-need-to-learn-how-to-read-it

Connuck, J., ed. (2022). "Offsetted." In *Cooking Sections: Offsetted* (pp. 17–41). Hatje Cantz Verlag.

Cosgrove, D. (1985). "Prospect, Perspective and the Evolution of the Landscape Idea." *Transactions of the Institute of British Geographers*, *10*, 45–62.

Cronon, W. (1991). *Nature's Metropolis: Chicago and the Great West*. W. W. Norton.

Crosby, A. W. (1972). *The Columbian Exchange: Biological and Cultural Consequences of 1492*. Greenwood.

Croucher, P. J. P., Mascheretti, S., & Garbelotto, M. (2013). "Combining Field Epidemiological Information and Genetic Data to Comprehensively Reconstruct the Invasion History and the Microevolution of the Sudden Oak Death Agent *Phytophthora Ramorum* (Stramenopila: Oomycetes) in California."

Biological Invasions, 15(10), 2281–2297. https://doi.org/10.1007/s10530-013-0453-8

Dahm, C. N., Candelaria-Ley, R. I., Reale, C. S., Reale, J. K., & Van Horn, D. J. (2015). "Extreme Water Quality Degradation Following a Catastrophic Forest Fire." *Freshwater Biology, 60*(12), 2584–2599. https://doi.org/10.1111/fwb.12548

Dambrauskaite, M. (2019). *Incidence of Heterobasidion spp. in Scots Pine on "Low Risk" Sites in Southern Sweden.* Master thesis. Swedish University of Agricultural Science (SLU). https://stud.epsilon.slu.se/14672/11/__ad.slu.se_common_bibul_slub_Arkiv_AVD_Vet_Kom_Publicering_epsilon_examensarbeten_examensarbeten19_Dambrauskaite_M_190703.pdf

Davis, H., & Todd, Z. (2017). "On the Importance of a Date, or Decolonizing the Anthropocene." *ACME: An International Journal for Critical Geographies, 16*, 761–780.

Davis, H., & Turpin, E., eds. (2015). *Art in the Anthropocene: Encounters Among Aesthetics, Politics, Environments and Epistemologies.* Open Humanities Press. https://doi.org/10.26530/OAPEN_560010

Davis, J., Moulton, A., Van Sant, L., & Williams, B. (2019). "Anthropocene, Capitalocene, . . . Plantationocene?: A Manifesto for Ecological Justice in an Age of Global Crises." *Geography Compass, 13*(5), Article 5. https://doi.org/10.1111/gec3.12438

de la Cadena, M. (2015). *Earth Beings: Ecologies of Practice Across Andean Worlds.* Duke University Press.

Depotter, J. R., Seidl, M. F., Wood, T. A., & Thomma, B. P. (2016). "Interspecific Hybridization Impacts Host Range and Pathogenicity of Filamentous Microbes." *Current Opinion in Microbiology, 32*, 7–13. https://doi.org/10.1016/j.mib.2016.04.005

de Ribeiro, P., Sujii, E. R., Diniz, I. R., de Medeiros, M. A., Salgado-Labouriau, M. L., Branco, M. C., Pires, C. S. S., & Fontes, E. M. G. (2010). "Alternative Food Sources and Overwintering Feeding Behavior of the Boll Weevil, *Anthonomus grandis* boheman (Coleoptera: Curculionidae) Under the Tropical Conditions of Central Brazil." *Neotropical Entomology, 39*(1), 28–34. https://doi.org/10.1590/S1519-566X2010000100005

Desimini, J. (2019). "Cartographic Grounds: The Temporal Cases." In *Mapping Landscapes in Transformation*, edited by T. Coomans, B. Cattoor, & K. De Jonge (pp. 17–43). Leuven University Press. https://doi.org/10.11116/9789461662835

De Sy, V., Herold, M., Achard, F., Beuchle, R., Clevers, J. G. P. W., Lindquist, E., & Verchot, L. (2015). "Land Use Patterns and Related Carbon Losses Following Deforestation in South America." *Environmental Research Letters, 10*(12), 124004. https://doi.org/10.1088/1748-9326/10/12/124004

Devlieger, L. (2020). "Reverse Architecture. The Virtues of Unbuilding and Reassembling." In *Rewriting Architecture: 10+1 Actions: Tabula Scripta*, edited by F. Alkemade, M. Minkjan, M. van Iersel, & J. Ouburg (pp. 147–152). Valiz.

———. (2021). "Once in German Rivers, *Eriocheir sinensis* Spread Fast." In

Feral Atlas: The More-Than-Human Anthropocene, edited by A. L. Tsing, J. Deger, A. K. Saxena, & F. Zhou. Stanford University Press. https://feralatlas.supdigital.org/poster/once-in-german-rivers-eriocheir-sinensis-spread-fast

Dewan, C. 2021. *Misreading the Bengal Delta: Climate Change, Development, and Livelihoods in Coastal Bangladesh.* University of Washington Press.

Dickens, J. C. (1986). "Orientation of Boll Weevil, *Anthonomus grandis* boh. (Coleoptera: Curculionidae), to Pheromone and Volatile Host Compound in the Laboratory." *Journal of Chemical Ecology,* 12(1), 91–98. https://doi.org/10.1007/BF01045593

Diouf, S. A. (2014). *Slavery's Exiles: The Story of the American Maroons.* New York University Press.

Doherty, J. (2021). "Uninvited Guests Take Advantage of the Feast Available in the City." In *Feral Atlas: The More-Than-Human Anthropocene,* edited by A. L. Tsing, J. Deger, A. K. Saxena, & F. Zhou. Stanford University Press. https://feralatlas.supdigital.org/poster/uninvited-guests-take-advantage-of-the-feast-available-to-them-in-the-city

Doody, J. S., Soennichsen, K. F., James, H., McHenry, C., & Clulow, S. (2021). "Ecosystem Engineering by Deep-Nesting Monitor Lizards." *Ecology, 102*(4), e03271. https://doi.org/10.1002/ecy.3271

Doyle, J. (1985). *Altered Harvest: Agriculture, Genetics, and the Fate of the World's Food Supply.* Viking.

Drummond-Cole, A. (2021). "What Happens to Lost Water?" In *Feral Atlas: The More-Than-Human Anthropocene,* edited by A. L. Tsing, J. Deger, A. K. Saxena, & F. Zhou. Stanford University Press. https://feralatlas.supdigital.org/poster/what-happens-to-lost-water

du Plessis, P. (2018). *Gathering the Kalahari: Tracking Landscapes in Motion.* PhD dissertation. Department of Anthropology, University of California, Santa Cruz.

———. (2021). "When Boreholes Become Herders, Cattle Menace Grasslands." In *Feral Atlas: The More-Than-Human Anthropocene,* edited by A. L. Tsing, J. Deger, A. K. Saxena, & F. Zhou. Stanford University Press. https://feralatlas.supdigital.org/poster/when-boreholes-become-herders-cattle-menace-grasslands

Dupré, G., & Rey, P.-P. (1973). "Reflections on the Pertinence of a Theory of the History of Exchange." *Economy and Society,* 2(2), 131–163. https://doi.org/10.1080/03085147300000007

Ebron, P. A. (2021). "Slave Ships Were Incubators for Infectious Diseases." In *Feral Atlas: The More-Than-Human Anthropocene,* edited by A. L. Tsing, J. Deger, A. K. Saxena, & F. Zhou. Stanford University Press. https://feralatlas.supdigital.org/poster/slave-ships-were-incubators-for-infectious-diseases

Ellis, E. C., Goldewijk, K. K., Siebert, S., Lightman, D., & Ramankutty, N. (2010). "Anthropogenic Transformation of the Biomes, 1700 to 2000." *Global Ecology and Biogeography,* 19, 589–606.

EPA (U.S. Environmental Protection Agency). (2021). "Chinook Salmon." *US EPA.* https://www.epa.gov/salish-sea/chinook-salmon

Evans-Pritchard, E. E. (1950). "Social Anthropology: Past and Present: The Marett Lecture, 1950." *Man*, *50*, 118–124. https://doi.org/10.2307/2794464

Everson, A. (2004). *Heritage* [Giclée Print].

Fach, B., Salihoglu, B., & Oğuz, T. (2021). "Alien Species Can Cause Severe Disturbance." In *Feral Atlas: The More-Than-Human Anthropocene*, edited by A. L. Tsing, J. Deger, A. K. Saxena, & F. Zhou. Stanford University Press. https://feralatlas.supdigital.org/poster/the-invasion-of-alien-species-can-cause-severe-disturbance-and-regime-shifts-to-ecosystems

Farrer, R. A., Weinert, L. A., Bielby, J., Garner, T. W. J., Balloux, F., Clare, F., Bosch, J., Cunningham, A. A., Weldon, C., du Preez, L. H., Anderson, L., Pond, S. L. K., Shahar-Golan, R., Henk, D. A., & Fisher, M. C. (2011). "Multiple Emergences of Genetically Diverse Amphibian-Infecting Chytrids Include a Globalized Hypervirulent Recombinant Lineage." *PNAS, 108*(46), 18732–18736. https://doi.org/10.1073/pnas.1111915108

Fearnley, L., & Lynteris, C. (2021). "Would Shutting Down China's Wet Markets Be a Big Mistake?" In *Feral Atlas: The More-Than-Human Anthropocene*, edited by A. L. Tsing, J. Deger, A. K. Saxena, & F. Zhou. Stanford University Press. https://feralatlas.supdigital.org/poster/coronavirus-stories-are-still-emerging

Fenn, E. A. (2002). *Pox Americana: The Great Smallpox Epidemic of 1775–82*. Macmillan.

———. (2014). *Encounters at the Heart of the World: A History of the Mandan People*. Farrar, Straus and Giroux.

———. (2021). "Grandfather Snake Waits Patiently for a River to Run Free." In *Feral Atlas: The More-Than-Human Anthropocene*, edited by A. L. Tsing, J. Deger, A. K. Saxena, & F. Zhou. Stanford University Press. https://feralatlas.supdigital.org/poster/grandfather-snake-waits-patiently-for-a-river-to-run-free

Fennelly, B. A. (2008). "The Kudzu Chronicles." In *Unmentionables: Poems* (pp. 61–80). W. W. Norton.

Ficek, R. E. (2021). "Pasture Grasses Are Barriers to Forest Regeneration in Latin America." In *Feral Atlas: The More-Than-Human Anthropocene*, edited by A. L. Tsing, J. Deger, A. K. Saxena, & F. Zhou. Stanford University Press. https://feralatlas.supdigital.org/poster/pasture-grasses-are-major-barriers-to-forest-regeneration-in-latin-america

Fisher, M. C., & Garner, T. W. J. (2007). "The Relationship Between the Emergence of *Batrachochytrium dendrobatidis*, the International Trade in Amphibians and Introduced Amphibian Species." *Fungal Biology Reviews*, *21*(1), 2–9. https://doi.org/10.1016/j.fbr.2007.02.002

Fisher, M. C., Garner, T. W. J., & Walker, S. F. (2009). "Global Emergence of *Batrachochytrium dendrobatidis* and Amphibian Chytridiomycosis in Space, Time, and Host." *Annual Review of Microbiology*, *63*(1), 291–310. https://doi.org/10.1146/annurev.micro.091208.073435

Fisher, M. C., Gurr, S. J., Cuomo, C. A., Blehert, D. S., Jin, H., Stukenbrock, E. H., Stajich, J. E., Kahmann, R., Boone, C., Denning, D. W., Gow, N. A. R., Klein, B. S., Kronstad, J. W., Sheppard, D. C., Taylor, J. W., Wright, G. D.,

Heitman, J., Casadevall, A., & Cowen, L. E. (2020). "Threats Posed by the Fungal Kingdom to Humans, Wildlife, and Agriculture." *mBio*, *11*(3), e00449–20. https://doi.org/10.1128/mBio.00449-20
Fisher, M. C., Henk, Daniel. A., Briggs, C. J., Brownstein, J. S., Madoff, L. C., McCraw, S. L., & Gurr, S. J. (2012). "Emerging Fungal Threats to Animal, Plant and Ecosystem Health." *Nature*, 484(7393), 186–194. https://doi.org/10.1038/nature10947
Flecker, A. S., Twining, C. W., Schmitz, O. J., Cooke, S. J., & Hammerschlag, N. (2019). "Aquatic Predators Influence Micronutrients: Important but Understudied." *Trends in Ecology & Evolution*, 34(10), 882–883. https://doi.org/10.1016/j.tree.2019.07.006
Fonseca, F. (2022a, April 22). "Sheriff's Office: Flagstaff-Area Wildfire Burned 30 Homes." *Arizona Republic*. https://www.azcentral.com/story/news/local/arizona/2022/04/22/sheriffs-office-flagstaff-tunnel-wildfire-burned-30-homes/7410114001/
———. (2022b, June 20). "Arizona Wildfires Sweep Lands Rich with Ancient Sites, Artifacts." *PBS NewsHour*. https://www.pbs.org/newshour/nation/arizona-wildfires-sweep-lands-rich-with-ancient-sites-artifacts
Fonseca, F., & Sullivan, T. (2020, April 11). "'The Grief Is So Unbearable': Virus Takes Toll on Navajo." *AP NEWS*. https://apnews.com/article/arizona-nm-state-wire-lifestyle-native-americans-understanding-the-outbreak-c77cc3c537c9a2510b67dcb631b4d988
Forseth, I. N., & Innis, A. F. (2021). "Kudzu Can Be a Major Ecosystem Threat." In *Feral Atlas: The More-Than-Human Anthropocene*, edited by A. L. Tsing, J. Deger, A. K. Saxena, & F. Zhou. Stanford University Press. https://feralatlas.supdigital.org/poster/kudzu-pueraria-montana-history-physiology-and-ecology-combine-to-make-a-major-ecosystem-threat
Foster, D. R., ed. (2014). *Hemlock: A Forest Giant on the Edge*, co-edited by A. D'Amato, B. Baiser, A. M. Ellison, D. Foster, D. Orwig, W. Oswald, A. B. Plotkin, & J. Thompson; consulting ed. S. Long. Yale University Press.
Francaviglia, R. V. (1978). *The Mormon Landscape: Existence, Creation, and Perception of a Unique Image in the American West*. AMS Press.
Freidberg, S. (2010). *Fresh: A Perishable History*. Belknap Press.
Frickel, S. (2021). "Cities Are Subterranean Disasters." In *Feral Atlas: The More-Than-Human Anthropocene*, edited by A. L. Tsing, J. Deger, A. K. Saxena, & F. Zhou. Stanford University Press. https://feralatlas.supdigital.org/poster/cities-are-subterranean-disasters
Friz, A., & Zunino, R. R. (2021). "Waste Rock Piles from Mining Shape the Desert for Millennia to Come." In *Feral Atlas: The More-Than-Human Anthropocene*, edited by A. L. Tsing, J. Deger, A. K. Saxena, & F. Zhou. Stanford University Press. https://feralatlas.supdigital.org/poster/overburden
Fujikane, C. (2021). *Mapping Abundance for a Planetary Future: Kanaka Maoli and Critical Settler Cartographies in Hawai'i*. Duke University Press.
Fulmer, J. E. (2009). "What in the World Is Infrastructure?" *PEI Infrastructure Investor*, (July/August), 30–32.
Funch, P. (2021). "Is There a Trojan Horse in Drinking Water?" In *Feral Atlas:*

The More-Than-Human Anthropocene, edited by A. L. Tsing, J. Deger, A. K. Saxena, & F. Zhou. Stanford University Press. https://feralatlas.supdigital.org/poster/is-there-a-trojan-horse-in-drinking-water

Gabrys, J. (2021). "As Lichens Absorb Air Pollution, They Show Us the Limits of Livability." In *Feral Atlas: The More-Than-Human Anthropocene*, edited by A. L. Tsing, J. Deger, A. K. Saxena, & F. Zhou. Stanford University Press. https://feralatlas.supdigital.org/poster/as-lichens-absorb-air-pollution-they-show-us-the-limits-of-livability

Gan, E. (2021). "Brown Planthoppers Rule Green Revolution Rice Fields." In *Feral Atlas: The More-Than-Human Anthropocene*, edited by A. L. Tsing, J. Deger, A. K. Saxena, & F. Zhou. Stanford University Press. https://feralatlas.supdigital.org/poster/brown-planthopper-outbreaks-are-an-unprecedented-and-devastating-effect-of

Gan, E., & Tsing, A. (2018). "How Things Hold: A Diagram of Coordination in a Satoyama Forest." *Social Analysis*, *62*(4), 102–145. https://doi.org/10.3167/sa.2018.620406

Gan, E., Tsing, A., & Sullivan, D. (2018). "Using Natural History in the Study of Industrial Ruins." *Journal of Ethnobiology*, *38*(1), 39–54. https://doi.org/10.2993/0278-0771-38.1.039

Garbelotto, M. (2021). "Will Emergent Diseases Decimate Our Forests?" In *Feral Atlas: The More-Than-Human Anthropocene*, edited by A. L. Tsing, J. Deger, A. K. Saxena, & F. Zhou. Stanford University Press. https://feralatlas.supdigital.org/poster/will-emergent-diseases-decimate-our-forests

Garbelotto, M., & Gonthier, P. (2013). "Biology, Epidemiology, and Control of *Heterobasidion* Species Worldwide." *Annual Review of Phytopathology*, *51*(1), 39–59. https://doi.org/10.1146/annurev-phyto-082712-102225

Geenen, K. (2012). "How the People of Butembo (RDC) Were Chosen to Embody 'the New Congo': Or What the Appearance of a Poster in a City's Public Places Can Teach About Its Social Tissue." *International Journal of Urban and Regional Research*, *36*(3), 448–461. https://doi.org/10.1111/j.1468-2427.2011.01084.x

Geertz, C. (1963). *Agricultural Involution: The Processes of Ecological Change in Indonesia*. University of California Press.

Ghosh, A. (2021a). "Empire." In *Feral Atlas: The More-Than-Human Anthropocene*, edited by A. L. Tsing, J. Deger, A. K. Saxena, & F. Zhou. Stanford University Press. https://feralatlas.supdigital.org/index?text=ghosh-empire&ttype=essay&cd=true

———. (2021b). *The Nutmeg's Curse: Parables for a Planet in Crisis*. John Murray.

Gibson-Graham, J. K. (1996). *The End of Capitalism (As We Knew It): A Feminist Critique of Political Economy* (NED—New edition). University of Minnesota Press.

Gibson-Graham, J. K., & Miller, E. (2015). "Economy as Ecological Livelihood." In *Manifesto for Living in the Anthropocene*, edited by K. Gibson, D. B. Rose, & R. Fincher (pp. 7–16). Punctum Books.

Giesen, J. C. (2011). *Boll Weevil Blues: Cotton, Myth, and Power in the American South*. University of Chicago Press.

Gilbert, S. F. (2021). "Holobionts Can Evolve by Changing Their Symbionts and Hosts." In *Feral Atlas: The More-Than-Human Anthropocene*, edited by A. L. Tsing, J. Deger, A. K. Saxena, & F. Zhou. Stanford University Press. https://feralatlas.supdigital.org/poster/holobionts-can-evolve-by-changing-their-symbionts-and-hosts

Gilbert, S. F., & Epel, D. (2015). *Ecological Developmental Biology: The Environmental Regulation of Development, Health, and Evolution,* 2nd ed. Sinauer Associates.

Gilroy, P. (1995). *The Black Atlantic: Modernity and Double-Consciousness.* Harvard University Press.

Goodluck, K. (2020, May 19). "COVID-19 Impacts Every Corner of the Navajo Nation." *High Country News*. https://www.hcn.org/issues/52.6/covid-19-impacts-every-corner-of-the-navajo-nation

Gopal, B. (1987). *Water Hyacinth.* Elsevier.

Gordon, D. M. (2021). "The Effect of an Invasive Species Is Not Constant." In *Feral Atlas: The More-Than-Human Anthropocene*, edited by A. L. Tsing, J. Deger, A. K. Saxena, & F. Zhou. Stanford University Press. https://feralatlas.supdigital.org/poster/the-effect-of-an-invasive-species-is-not-constant

Graber, B. (2022, November 17). "Five Key Lessons as World's Biggest Dam Removal Project Will Soon Begin on the Klamath River." *American Rivers: River Stories.* https://www.americanrivers.org/2022/11/five-key-lessons-as-worlds-biggest-dam-removal-project-will-soon-begin-on-the-klamath-river/

Grandin, G. (2010). *Fordlandia: The Rise and Fall of Henry Ford's Forgotten Jungle City.* Henry Holt.

Grossberg, L. (1986). "On Postmodernism and Articulation: An Interview with Stuart Hall." *Journal of Communication Inquiry*, 10(2), 45–60. https://doi.org/10.1177/019685998601000204

Gurrumuruwuy, P., Deger, J., Guruŋulmiwuy, E., Balpatji, W., Balanydjarrk, M., & Coffey, V. B. (in press). "Luŋ'thun: Sand, Saltwater and Collaborative Attunements." *Australian Journal of Anthropology, Double Special Issue Epistemic Attunements: Regenerating Anthropology's Form.*

Hadfield, M. G. (2021). "Snails That Eat Snails." In *Feral Atlas: The More-Than-Human Anthropocene*, edited by A. L. Tsing, J. Deger, A. K. Saxena, & F. Zhou. Stanford University Press. https://feralatlas.supdigital.org/poster/snails-that-eat-snails

Hall, S. (2018). "Race, Articulation, and Societies Structured in Dominance" [1980]. In *Essential Essays, Volume 1*, edited by D. Morley (pp. 172–221). Duke University Press. https://doi.org/10.1215/9781478002413-010

Hammerschlag, N., Schmitz, O. J., Flecker, A. S., Lafferty, K. D., Sih, A., Atwood, T. B., Gallagher, A. J., Irschick, D. J., Skubel, R., & Cooke, S. J. (2019). "Ecosystem Function and Services of Aquatic Predators in the Anthropocene." *Trends in Ecology & Evolution*, 34(4), 369–383. https://doi.org/10.1016/j.tree.2019.01.005

Haney, P., Lewis, J., & Lambert, W. (2009). "Cotton Production and the Boll Weevil in Georgia: History, Cost of Control, and Benefits of Eradication."

University of Georgia Research Bulletin, 428. http://athenaeum.libs.uga.edu/bitstream/handle/10724/12179/RB428.pdf?sequence=1

Haraway, D. (2008). *When Species Meet.* University of Minnesota Press.

———. (2016). *Staying with the Trouble: Making Kin in the Chthulucene.* Duke University Press.

Hardt, M., & Negri, A. (2003). *Empire.* Harvard University Press.

Hassell, J., Begon, M., Ward, M., & Fèvre, E. (2021). "In Urban Environmental Interfaces, Humans, Domestic Animals, and Wildlife Interact Creating Disease Emergence Opportunities." In *Feral Atlas: The More-Than-Human Anthropocene*, edited by A. L. Tsing, J. Deger, A. K. Saxena, & F. Zhou. Stanford University Press. https://feralatlas.supdigital.org/poster/coronavirus-stories-are-still-emerging

Haus der Kulturen der Welt and the Max Planck Institute for the History of Science. (2018). *Mississippi An Anthropocene River 2018–19.* https://www.hkw.de/en/programm/projekte/2018/mississippi_an_anthropocene_river/ _an_anthropocene_river_start.php

Hausdoerffer, J., Kimmerer, R. W., Blakie, S., Salmón, E., Williams, O., & Morales, M. I. (2021). "Epilogue—Attention, Curiosity, Play, Gratitude: Practices of Kinship." In *Kinship: Belonging in a World of Relations: Volume 5: Practice,* edited by G. Van Horn, R. W. Kimmerer, & J. Hausdoerffer (pp. 127–147). Center for Humans and Nature.

Hayakawa, Y. (2013). "Radioactive Contamination Map (8th Edition) Monochrome Simplified Specification." *Yukio Hayakawa's Volcano Blog.* http://kipuka.blog70.fc2.com/blog-category-20.html

Hayes, K. A., Joshi, R. C., Thiengo, S. C., & Cowie, R. H. (2008). "Out of South America: Multiple Origins of Non-Native Apple Snails in Asia." *Diversity and Distributions*, *14*(4), 701–712. https://doi.org/10.1111/j.1472-4642.2008.00483.x

Hazen, R., Papineau, D., Bleeker, W., Downs, R., Ferry, J. M., McCoy, T., Sverjensky, D., & Yang, H. (2012). "Mineral Evolution." *American Mineralogist, 93*, 1693–1720.

Hébert, K. (2015). "Enduring Capitalism: Instability, Precariousness, and Cycles of Change in an Alaskan Salmon Fishery." *American Anthropologist, 117*(1), 32–46. https://doi.org/10.1111/aman.12172

Hecht, G. (2021). "No Need to Wash the Goose." In *Feral Atlas: The More-Than-Human Anthropocene,* edited by A. L. Tsing, J. Deger, A. K. Saxena, & F. Zhou. Stanford University Press. https://feralatlas.supdigital.org/index?text=gabrielle-hecht-no-need-to-wash-the-goose-a-meditation-on-waste&ttype=essay&cd=true

Heglar, M. A. (2019, February 18). "Climate Change Isn't the First Existential Threat." *Medium.* https://zora.medium.com/sorry-yall-but-climate-change-ain-t-the-first-existential-threat-b3c999267aa0

Heiddeger, M. (1995). *The Fundamental Concepts of Metaphysics: World, Finitude, Solitude,* translated by W. McNeill & N. Walker. Indiana University Press.

Hesse-Honegger, C. (2021). "Downwind . . ." In *Feral Atlas: The More-Than-*

Human Anthropocene, edited by A. L. Tsing, J. Deger, A. K. Saxena, & F. Zhou. Stanford University Press. https://feralatlas.supdigital.org/poster/downwind

Ho, K. (2021). "Finance, Circa 1980s: The Acceleration of Extraction." In *Feral Atlas: The More-Than-Human Anthropocene*, edited by A. L. Tsing, J. Deger, A. K. Saxena, & F. Zhou. Stanford University Press. https://feralatlas.supdigital.org/index?text=ho-acceleration&ttype=essay&cd=true

Hodgetts, T., & Lorimer, J. (2020). "Animals' Mobilities." *Progress in Human Geography*, 44(1), 4–26, Article 1. https://doi.org/10.1177/0309132518817829

Hoff, R. J., Ferguson, D. E., McDonald, G. I., & Keane, R. E. (2001). "Strategies for Managing Whitebark Pine in the Presence of White Pine Blister Rust." In *Whitebark Pine Communities: Ecology and Restoration*, edited by D. F. Tomback, S. F. Arno, & R. E. Keane (pp. 346–366). Island Press.

Holbraad, M., & Pedersen, M. (2017). *The Ontological Turn*. Cambridge University Press.

Holmes, D., Murray, S., Perron, A., & Rail, G. (2006). "Deconstructing the Evidence-Based Discourse in Health Sciences: Truth, Power and Fascism." *International Journal of Evidence-Based Healthcare*, 4, 180–186.

Howes, D. (2013, August). "The Expanded Field of Sensory Studies." *Sensory Studies*. https://www.sensorystudies.org/sensorial-investigations/the-expanding-field-of-sensory-studies/

hudsonriverkeeper, dir. (2021, October 4). *Habitat Restoration Along the Hudson—Riverkeeper*. https://www.youtube.com/watch?v=toQgUnyEU6U

Hulme, M. (2011). "Reducing the Future to Climate: A Story of Climate Determinism and Reductionism." *Osiris*, 26(1), 245–266. https://doi.org/10.1086/661274

Ingold, T. (2021). "Globes and Spheres: The Topology of Environmentalism." In *The Perception of the Environment: Essays on Livelihood, Dwelling and Skill*. Routledge.

Institute of Medicine. (2011). *Fungal Diseases: An Emerging Threat to Human, Animal, and Plant Health: Workshop Summary*. National Academies Press. https://doi.org/10.17226/13147

Iqbal, I. (2010). *The Bengal Delta: Ecology, State and Social Change, 1840–1943*. Palgrave Macmillan.

———. (2021). "In the Bengal Delta, the Anthropocene Began with the Arrival of the Railways." In *Feral Atlas: The More-than-Human Anthropocene*, edited by A. L. Tsing, J. Deger, A. K. Saxena, & F. Zhou. Stanford University Press. https://feralatlas.supdigital.org/poster/in-the-bengal-delta-the-anthropocene-began-with-the-arrival-of-the-railways

Isaac, S. (1992). *Fungal–Plant Interactions*. Chapman and Hall.

Ishida, N., & Naito, D. (2021). "Fukushima Waste Ends Up in Public Works Projects." In *Feral Atlas: The More-Than-Human Anthropocene*, edited by A. L. Tsing, J. Deger, A. K. Saxena, & F. Zhou. Stanford University Press. https://feralatlas.supdigital.org/poster/public-works-distribute-fukushima-waste

Jones, A., & Jenkins, K. (2014). "Rethinking Collaboration: Working the Indigene-Colonizer Hyphen." In *Handbook of Critical and Indigenous Method-*

ologies, edited by N. K. Denzin, Y. S. L. Lincoln, & L. T. Smith (pp. 471–486). Sage Publications. https://doi.org/10.4135/9781483385686.n23

Jones, R. W. (2001). "Evolution of the Host Plant Associations of the *Anthonomus grandis* Species Group (Coleoptera: Curculionidae): Phylogenetic Tests of Various Hypotheses." *Annals of the Entomological Society of America*, 94(1), 51–58. https://doi.org/10.1603/0013-8746(2001)094[0051:EOTHPA]2.0.CO;2

Jordan, C. (2021). "This Is Our Culture Turned Inside Out." In *Feral Atlas: The More-Than-Human Anthropocene*, edited by A. L. Tsing, J. Deger, A. K. Saxena, & F. Zhou. Stanford University Press. https://feralatlas.supdigital.org/poster/this-is-our-culture-turned-inside-out

Joshi, R., & Sebastian, L. (2006). *Global Advances in Ecology and Management of Golden Apple Snails*. Philippine Rice Research Institute.

Kadono, Y. (2004). "Alien Aquatic Plants Naturalized in Japan: History and Present Status." *Global Environmental Research*, 8(2), 163–169.

Kauserud, H., Svegården, I. B., Sætre, G.-P., Knudsen, H., Stensrud, Ø., Schmidt, O., Doi, S., Sugiyama, T., & Högberg, N. (2007). "Asian Origin and Rapid Global Spread of the Destructive Dry Rot Fungus *Serpula lacrymans*." *Molecular Ecology*, 16(16), 3350–3360. https://doi.org/10.1111/j.1365-294X.2007.03387.x

Keane, W. (2008). "Market, Materiality and Moral Metalanguage." *Anthropological Theory*, 8(1), 27–42. https://doi.org/10.1177/1463499607087493

Keck, F. (2021). "Museum Visitors Carry Clouds of Insects." In *Feral Atlas: The More-Than-Human Anthropocene*, edited by A. L. Tsing, J. Deger, A. K. Saxena, & F. Zhou. Stanford University Press. https://feralatlas.supdigital.org/poster/museum-visitors-carry-clouds-of-insects

Keleman, A. (2005). "Methodology in a Time of Crackdown." *Tropical Resources: The Bulletin of the Yale Tropical Resources Institute*, 24, 9–14.

———. (2010). "Institutional Support and In Situ Conservation in Mexico: Biases Against Small-Scale Maize Farmers in Post-NAFTA Agricultural Policy." *Agriculture and Human Values*, 27(1), 13–28. https://doi.org/10.1007/s10460-009-9192-y

Keleman Saxena, A. (2021a). "A View from a Patch: Toward a Material Phenomenology of Climate Change." In *Feral Atlas: The More-than-Human Anthropocene*, edited by A. L. Tsing, J. Deger, A. K. Saxena, & F. Zhou. Stanford University Press. https://feralatlas.supdigital.org/index?text=alder-keleman-saxena-patchy-climate-change&ttype=essay&cd=true

———. (2021b). "In the High Andes, Roads Encourage Disease and Science." In *Feral Atlas: The More-than-Human Anthropocene*, edited by A. L. Tsing, J. Deger, A. K. Saxena, & F. Zhou. Stanford University Press. https://feralatlas.supdigital.org/poster/in-the-high-altitude-andes-native-potato-diversity-is-newly-vulnerable-to

Keleman Saxena, A., Cadima Fuentes, X., Gonzales Herbas, R., & Humphries, D. L. (2016). "Indigenous Food Systems and Climate Change: Impacts of Climatic Shifts on the Production and Processing of Native and Traditional

Crops in the Bolivian Andes." *Frontiers in Public Health*, 4. https://doi.org/10.3389/fpubh.2016.00020

Keogh, L. (2011). "Introducing the Cane Toad." *Queensland Historical Atlas*. https://www.qhatlas.com.au/introducing-cane-toad

Kimmerer, R. W. (2014). *Braiding Sweetgrass: Indigenous Wisdom, Scientific Knowledge, and the Teachings of Plants*. Milkweed Editions.

Kitchin, R., & Dodge, M. (2007). "Rethinking Maps." *Progress in Human Geography*, *31*(3), 331–344, Article 3. https://doi.org/10.1177/0309132507077082

Koch, A., Brierley, C., Maslin, M. A., & Lewis, S. L. (2019). "Earth System Impacts of the European Arrival and Great Dying in the Americas After 1492." *Quaternary Science Reviews*, 207, 13–36.

Kohn, E. (2013). *How Forests Think: Toward an Anthropology Beyond the Human*. University of California Press.

———. (2015). "Anthropology of Ontologies." *Annual Review of Anthropology*, 44(1), 311–327. https://doi.org/10.1146/annurev-anthro-102214-014127

Kolby, J., & Berger, L. (2021). "Global Trade in Frogs Has Led to Catastrophic Amphibian Decline." In *Feral Atlas: The More-Than-Human Anthropocene*, edited by A. L. Tsing, J. Deger, A. K. Saxena, & F. Zhou. Stanford University Press. https://feralatlas.supdigital.org/poster/global-trade-in-frogs-has-led-to-disease-spread-and-catastrophic-amphibian

Kolby, J. E. (2014). "Stop Madagascar's Toad Invasion Now." *Nature*, *509*(7502), 563–563. https://doi.org/10.1038/509563a

Korotaev, G., Oguz, T., Nikiforov, A., & Koblinsky, C. (2003). "Seasonal, Interannual, and Mesoscale Variability of the Black Sea Upper Layer Circulation Derived from Altimeter Data." *Journal of Geophysical Research: Oceans*, *108*(C4). https://doi.org/10.1029/2002JC001508

Kowalewska, A. (2021). "We Set Fire to the Sea. Toxic Glowing Algal Blooms in the Baltic." In *Feral Atlas: The More-Than-Human Anthropocene*, edited by A. L. Tsing, J. Deger, A. K. Saxena, & F. Zhou. Stanford University Press. https://feralatlas.supdigital.org/poster/we-set-fire-to-the-sea-toxic-glowing-algal-blooms-in-the-baltic

Lacaton, A., & Vassal, J.-P. (1996). *Place Léon Aucoc, Bordeaux*. https://www.lacatonvassal.com/index.php?idp=37#

Lacelli, G. A. (2004). *Cotton in Argentina: Evolution During Last Years and Actual Situation*. https://icac.org/Content/SEEPDocuments/PdfFilesd56dc678_eac5_4193_a0f3_cc0edf771e33/argentina.pdf

Ladyzhets, B., Montanari, S., & Monahan, R. (2022, December 28). "The Uncounted: People of Color Are Dying at Much Higher Rates Than What COVID Data Suggests." *University of Arizona Center for Rural Health: News*. https://crh.arizona.edu/news/uncounted-people-color-are-dying-much-higher-rates-what-covid-data-suggests

Lahoud, A. (2019, August 9). "ArtReview x Sharjah Architecture Triennial: *Ngurrara II*." *ArtReview*. https://artreview.com/online-2019-sharjah-triennial-1-adrian-lahoud/

Lahsen, M. (2005). "Seductive Simulations? Uncertainty Distribution Around

Climate Models." *Social Studies of Science*, 35(6), 895–922. https://doi.org/10.1177/0306312705053049
———. (2010). "The Social Status of Climate Change Knowledge: An Editorial Essay." *Interdisciplinary Reviews Climate Change*, 1(2), 162–171.
———. (2015). "Digging Deeper into the Why: Cultural Dimensions of Climate Change Skepticism Among Scientists." In *Climate Cultures: Anthropological Perspectives on Climate Change*, edited by J. Barnes & M. R. Dove (pp. 221–248). Yale University Press.
Lange, F., Olmstead, A. L., & Rhode, P. W. (2009). "The Impact of the Boll Weevil, 1892–1932." *Journal of Economic History*, 69(3), 685–718.
Larkin, B. (2013). "The Politics and Poetics of Infrastructure." *Annual Review of Anthropology*, 42(1), Article 1. https://doi.org/10.1146/annurev-anthro-092412-155522
Latour, B., & Weibel, P., eds. (2020). *Critical Zones: The Science and Politics of Landing on Earth*. MIT Press.
Law, J. (2015). "What's Wrong with a One-World World?" *Distinktion: Journal of Social Theory*, 16(1), 126–139. https://doi.org/10.1080/1600910X.2015.1020066
———. (2019). "Material Semiotics." *Heterogeneities*. www.heterogeneities.net/publications/Law2019MaterialSemiotics.pdf
Ledbetter, H. W. ("Lead Belly"). (2021). "He Was Looking for a Home." In *Feral Atlas: The More-Than-Human Anthropocene*, edited by A. L. Tsing, J. Deger, A. K. Saxena, & F. Zhou. Stanford University Press. https://feralatlas.supdigital.org/poster/he-was-looking-for-a-home
Le Guin, U. K. (1989). "The Carrier Bag Theory of Fiction." In *In Dancing at the End of the World* (pp. 165–170). Grove Press.
Lehrman, B. (2018). "Visualizing Water Infrastructure with Sankey Maps: A Case Study of Mapping the Los Angeles Aqueduct, California." *Journal of Maps*, 14(1), 52–64. https://doi.org/10.1080/17445647.2018.1473815
Leroux, S. J., & Schmitz, O. J. (2015). "Predator-Driven Elemental Cycling: The Impact of Predation and Risk Effects on Ecosystem Stoichiometry." *Ecology and Evolution*, 5(21), 4976–4988. https://doi.org/10.1002/ece3.1760
Lever, C. (2001). *The Cane Toad: The History and Ecology of a Successful Colonist*. Westbury Academic and Scientific Publishing.
Lewis, S. L., & Maslin, M. A. (2015). "Defining the Anthropocene." *Nature*, 519(7542), 171–180. https://doi.org/10.1038/nature14258
———. (2021). "The Great Dying of the Indigenous Peoples of the Americas." In *Feral Atlas: The More-Than-Human Anthropocene*, edited by A. L. Tsing, J. Deger, A. K. Saxena, & F. Zhou. Stanford University Press. https://feralatlas.supdigital.org/index?text=lewis-and-maslin-invasion&ttype=essay&cd=true
Li, T. M. (2017). "After Development: Surplus Population and the Politics of Entitlement." *Development and Change*, 48(6), 1247–1261, Article 6. https://doi.org/10.1111/dech.12344
Li, T. M., & Semedi, P. (2021). *Plantation Life: Corporate Occupation in Indonesia's Oil Palm Zone*. Duke University Press.

Lieberei, R. (2007). "South American Leaf Blight of the Rubber Tree (*Hevea* spp.): New Steps in Plant Domestication Using Physiological Features and Molecular Markers." *Annals of Botany*, *100*(6), 1125–1142. https://doi.org/10.1093/aob/mcm133

Liebman, A., & Wallace, R. G. (2021). "A Factory Farm Fungus Among Us." In *Feral Atlas: The More-Than-Human Anthropocene*, edited by A. Tsing, J. Deger, A. K. Saxena, & F. Zhou. Stanford University Press. https://feralatlas.supdigital.org/poster/a-factory-farm-fungus-among-us

Lien, A., & Camacho, E. (2021). "Artist Statement." In *Feral Atlas: The More-Than-Human Anthropocene.* edited by A. L. Tsing, J. Deger, A. K. Saxena, & F. Zhou. Stanford University Press. https://feralatlas.supdigital.org/index?text=amy-lien-and-enzo-camacho-artist-statement&ttype=essay&cd=true

Lien, M. E. (2015). *Becoming Salmon: Aquaculture and the Domestication of a Fish.* University of California Press.

Lin, A. T., & Myers, M. (2021). "Ancient Virus Genomes from Museum and Archaeological Collections Can Inform Past and Future Epidemics." In *Feral Atlas: The More-Than-Human Anthropocene*, edited by A. L. Tsing, J. Deger, A. K. Saxena, & F. Zhou. Stanford University Press. https://feralatlas.supdigital.org/poster/coronavirus-stories-are-still-emerging

Lin, M. (2009). *What Is Missing?* https://www.whatismissing.org/

Liu, A. B. (2021). "Is COVID-19 a 'Just-In-Time' Pandemic?" In *Feral Atlas: The More-Than-Human Anthropocene*, edited by A. L. Tsing, J. Deger, A. K. Saxena, & F. Zhou. Stanford University Press. https://feralatlas.supdigital.org/poster/is-covid-19-a-just-in-time-pandemic

Livingston, J. (2019). *Self-Devouring Growth: A Planetary Parable as Told from Southern Africa.* Duke University Press.

Ma, L.-J., van der Does, H. C., Borkovich, K. A., Coleman, J. J., Daboussi, M.-J., Di Pietro, A., Dufresne, M., Freitag, M., Grabherr, M., Henrissat, B., Houterman, P. M., Kang, S., Shim, W.-B., Woloshuk, C., Xie, X., Xu, J.-R., Antoniw, J., Baker, S. E., Bluhm, B. H., . . . Rep, M. (2010). "Comparative Genomics Reveals Mobile Pathogenicity Chromosomes in *Fusarium*." *Nature*, 464(7287), 367–373. https://doi.org/10.1038/nature08850

Macfarlane, R. (2015). *Landmarks.* Hamish Hamilton.

Mackenzie, D. (2021). "For Days We Circled the Mud." In *Feral Atlas: The More-Than-Human Anthropocene*, edited by A. L. Tsing, J. Deger, A. K. Saxena, & F. Zhou. Stanford University Press. https://feralatlas.supdigital.org/poster/for-days-we-circled-the-mud

Magrath, J., & Jennings, S. (2012). "What Happened to the Seasons? Farmers' Perceptions and Meteorological Observations of Changing Seasonality." In *Seasonality, Rural Livelihoods, and Development*, edited by S. Devereux, R. Sabates-Wheeler, & R. Longhurst. Earthscan.

Maguire, J. (2021). "Accelerated Energy Extraction Triggers Anthropogenic Earthquakes." In *Feral Atlas: The More-Than-Human Anthropocene*, edited by A. L. Tsing, J. Deger, A. K. Saxena, & F. Zhou. Stanford University Press. https://feralatlas.supdigital.org/poster/accelerated-energy-extraction-triggers-anthropogenic-earthquakes

Majnep, I. S., & Bulmer, R. (1977). *Birds of My Kalam Country*. University of Auckland University Press; Oxford University Press.

Malm, A., & Hornborg, A. (2014). "The Geology of Mankind? A Critique of the Anthropocene Narrative." *The Anthropocene Review*, *1*(1), Article 1. https://doi.org/10.1177/2053019613516291

Maloy, O. C. (1997). "White Pine Blister Rust Control in North America: A Case History." *Annual Review of Phytopathology*, *35*(1), 87–109. https://doi.org/10.1146/annurev.phyto.35.1.87

Malterre-Barthes, C. (2022). "A Global Moratorium on New Construction." https://www.charlottemalterrebarthes.com/practice/research-practice/a-global-moratorium-on-new-construction/

Marcon, F. (2015). *The Knowledge of Nature and the Nature of Knowledge in Early Modern Japan*. University of Chicago Press.

Marris, E. (2013). *Rambunctious Garden: Saving Nature in a Post-Wild World*. Bloomsbury.

Marx, K. (1992). *Capital, Volume 1: A Critique of Political Economy* [1887, English], edited by F. Engels. Penguin Classics.

———. (1994). *The Eighteenth Brumaire of Louis Bonaparte* [1869, English], translated by Daniel de Leon. International Publishers.

Massey, D. (2005). *For Space*. Sage Publications.

Mathews, A. S. (2018). "Landscapes and Throughscapes in Italian Forest Worlds: Thinking Dramatically About the Anthropocene." *Cultural Anthropology*, *33*(3), 386–414. https://doi.org/10.14506/ca33.3.05

———. (2021). "The Echoes of Exotic Diseases Are Visible in Italian Forests." In *Feral Atlas: The More-Than-Human Anthropocene*, edited by A. L. Tsing, J. Deger, A. K. Saxena, & F. Zhou. Stanford University Press. https://feralatlas.supdigital.org/poster/the-echoes-of-exotic-diseases-are-visible-in-italian-forests-if-we-know-how

Mauss, M. (1966). *The Gift: Forms and Functions of Exchange in Archaic Societies*. Cohen & West.

McCary, M. A., & Schmitz, O. J. (2021). "Invertebrate Functional Traits and Terrestrial Nutrient Cycling: Insights from a Global Meta-Analysis." *Journal of Animal Ecology*, *90*(7), 1714–1726. https://doi.org/10.1111/1365-2656.13489

McClintock, A. (2022, January). "Ghost Forest: Atlas of a Drowning World." *E-Flux Architecture*. https://www.e-flux.com/architecture/accumulation/440704/ghost-forest-atlas-of-a-drowning-world/

McCoy, K. A., & Peralta, A. L. (2018). "Pesticides Could Alter Amphibian Skin Microbiomes and the Effects of *Batrachochytrium dendrobatidis*." *Frontiers in Microbiology*, *9*. https://www.frontiersin.org/articles/10.3389/fmicb.2018.00748

McCully, P. (2001). *Silenced Rivers: The Ecology and Politics of Large Dams*. Zed Books.

McDinny, N. (2013). *Story of Mayawagu* [Acrylic on linen].

McNeill, J., & Engelke, P. (2014). *The Great Acceleration: An Environmental History of the Anthropocene Since 1945*. Belknap Press.

McNeill, J. R. (2010). *Mosquito Empires: Ecology and War in the Greater Caribbean, 1620–1914*. Cambridge University Press.

———. (2021). "Plantation Ecologies of the Caribbean Were Incubators for the Mosquitoes That Carry Two of Humankind's Most Lethal Diseases." In *Feral Atlas: The More-Than-Human Anthropocene*, edited by A. L. Tsing, J. Deger, A. K. Saxena, & F. Zhou. Stanford University Press. https://feralatlas.supdigital.org/poster/plantation-ecologies-of-the-caribbean-were-incubators-for-the-mosquitoes

Meillassoux, C. (1981). *Maidens, Meal, and Money: Capitalism and the Domestic Community*. Cambridge University Press.

Meis, J. F., Chowdhary, A., Rhodes, J. L., Fisher, M. C., & Verweij, P. E. (2016). "Clinical Implications of Globally Emerging Azole Resistance in *Aspergillus fumigatus*." *Philosophical Transactions of the Royal Society B: Biological Sciences*, 371(1709), 20150460. https://doi.org/10.1098/rstb.2015.0460

Miller, D., ed. (2005). *Materiality*. Duke University Press.

Milon, A.-S., & Zalasiewicz, J. (2021). "The Victims of Carbon Dioxide Are Starting to Appear." In *Feral Atlas: The More-Than-Human Anthropocene*, edited by A. L. Tsing, J. Deger, A. K. Saxena, & F. Zhou. Stanford University Press. https://feralatlas.supdigital.org/poster/the-victims-of-carbon-dioxide-are-starting-to-appear

Mintz, S. W. (1960). *Worker in the Cane: A Puerto Rican Life History*. Yale University Press.

———. (1986). *Sweetness and Power: The Place of Sugar in Modern History*. Penguin Books.

Mishuana, G. (2013). *Mark My Words: Native Women Mapping Our Nations*. University of Minnesota Press.

Misrach, R., & Orff, K. (2012). *Petrochemical America*. Aperture.

Miyarrka Media (Paul Gurrumuruwuy, Jennifer Deger, Enid Guruŋulmiwuy, Warren Balpatji, Meredith Balanydjarrk, James Ganambarr, & Kayleen Djingadjingawuy). (2019). *Phone & Spear: A Yuṯa Anthropology*. Goldsmiths Press.

Miyazaki, H., dir. (1984, March 11). *Nausicaä of the Valley of the Wind*. Toei Company. https://en.wikipedia.org/w/index.php?title=Nausica%C3%A4_of_the_Valley_of_the_Wind_(film)&oldid=1147637869

Mol, A. (2007). *The Body Multiple: Ontology in Medical Practice*. Duke University Press.

Monbiot, G. (2014). *Feral: Rewilding the Land, the Sea, and Human Life*. University of Chicago Press.

Moore, A. C., & Schmitz, O. J. (2021). "Do Predators Have a Role to Play in Wetland Ecosystem Functioning? An Experimental Study in New England Salt Marshes." *Ecology and Evolution*, 11(16), 10956–10967. https://doi.org/10.1002/ece3.7880

Moore, J. W. (2015). *Capitalism in the Web of Life: Ecology and the Accumulation of Capital*. Verso Books.

———, ed. (2016). *Anthropocene or Capitalocene? Nature, History, and the Crisis of Capitalism*. PM Press.
Morita, A. (2016a). "Infrastructuring Amphibious Space: The Interplay of Aquatic and Terrestrial Infrastructures in the Chao Phraya Delta in Thailand." *Science as Culture*, *25*(1), 117–140. https://doi.org/10.1080/09505431.2015.1081502
———. (2016b). "Traveling Within the Case." *Somatosphere*. http://somatosphere.net/2016/traveling-within-the-case.html/
———. (2017). "Multispecies Infrastructure: Infrastructural Inversion and Involutionary Entanglements in the Chao Phraya Delta, Thailand." *Ethnos*, *82*(4), 738–757. https://doi.org/10.1080/00141844.2015.1119175
Morita, A., & Jensen, C. B. (2017). "Delta Ontologies: Infrastructural Transformations in the Chao Phraya Delta, Thailand." *Social Analysis*, *61*(2), 118–133. https://doi.org/10.3167/sa.2017.610208
Morton, T. (2013). *Hyperobjects: Philosophy and Ecology After the End of the World*. University of Minnesota Press.
Mounzer, L. (2021). "Letter from Beirut: From Revolution to Pandemic." In *Feral Atlas: The More-Than-Human Anthropocene*, edited by A. L. Tsing, J. Deger, A. K. Saxena, & F. Zhou. Stanford University Press. https://feralatlas.supdigital.org/poster/letter-from-beirut-from-revolution-to-pandemic
Münster, U. (2021). "Lantana Invades Teak Plantations and Turns Elephants Violent." In *Feral Atlas: The More-Than-Human Anthropocene*, edited by A. L. Tsing, J. Deger, A. K. Saxena, & F. Zhou. Stanford University Press. https://feralatlas.supdigital.org/poster/lantana-invades-monocrop-teak-plantations-triggering-elephant-aggression
Murdock, G. (1957). "World Ethnographic Sample." *American Anthropologist*, *59*, 664–687.
Murrell, C., Gerber, E., Krebs, C., Parepa, M., Schaffner, U., & Bossdorf, O. (2011). "Invasive Knotweed Affects Native Plants Through Allelopathy." *American Journal of Botany*, *98*(1), 38–43. https://doi.org/10.3732/ajb.1000135
Nagy, K. (2021). "Plastics Saturate Us, Inside and Out." In *Feral Atlas: The More-Than-Human Anthropocene*, edited by A. L. Tsing, J. Deger, A. K. Saxena, & F. Zhou. Stanford University Press. https://feralatlas.supdigital.org/poster/plastics-saturate-us-inside-and-outside-our-bodies
Nash, L. (2006). *Inescapable Ecologies: A History of Environment, Disease, and Knowledge*. University of California Press.
Natural Resource Council of Maine. (2023). "Penobscot River Restoration Project." *Natural Resources Council of Maine*. https://www.nrcm.org/programs/waters/penobscot-river-restoration-project/
Navajo Nation Government. (2023, March 30). *COVID-19*. Navajo Department of Health. https://ndoh.navajo-nsn.gov/COVID-19
Nazarea, V. D. (1999). "A View from a Point: Ethnoecology as Situated Knowledge." In *Ethnoecology: Situated Knowledge/Located Lives*, edited by V. D. Nazarea. University of Arizona Press.
"Ngurrara, The Great Sandy Desert Canvas." (2008, March 5). *Aboriginal Art*

Directory. https://aboriginalartdirectory.com/ngurrara-the-great-sandy-desert-canvas/

Nnadi, N. E., & Carter, D. A. (2021). "Climate Change and the Emergence of Fungal Pathogens." *PLOS Pathogens*, *17*(4), e1009503. https://doi.org/10.1371/journal.ppat.1009503

Nordgren Ballivián, M. (2011). *Cambios Climáticos: Percepciones, efectos, y respuestas en cuatro regiones de Bolivia*. Centro de Investigación y Promoción del Campesinado (CIPCA).

Norton, R. J. (2018). "Feral Cities." *Naval War College Review*, *56*(4). https://digital-commons.usnwc.edu/nwc-review/vol56/iss4/8

Oda, F., Guerra Batista, V., Grou, E., Lima, L., Proença, H., Gambale, P., Takemoto, R., Pinheiro Teixeira, C., Campião, K., & Ortega, J. C. (2019). "Native Anuran Species as Prey of Invasive American Bullfrog *Lithobates catesbeianus* in Brazil: A Review with New Predation Records." *Amphibian and Reptile Conservation*, *13*, 217–226.

O'Hanlon, S. J., Rieux, A., Farrer, R. A., Rosa, G. M., Waldman, B., Bataille, A., Kosch, T. A., Murray, K. A., Brankovics, B., Fumagalli, M., Martin, M. D., Wales, N., Alvarado-Rybak, M., Bates, K. A., Berger, L., Böll, S., Brookes, L., Clare, F., Courtois, E. A., . . . Fisher, M. C. (2018). "Recent Asian Origin of Chytrid Fungi Causing Global Amphibian Declines." *Science*, *360*(6389), 621–627. https://doi.org/10.1126/science.aar1965

O'Loughlin, E., & Zaveri, M. (2020, May 5). "Irish Return an Old Favor, Helping Native Americans Battling the Virus." *New York Times*. https://www.nytimes.com/2020/05/05/world/coronavirus-ireland-native-american-tribes.html

Oreskes, N., & Conway, E. M. (2014). *The Collapse of Western Civilization: A View from the Future*. Columbia University Press.

Otto, T., & Bubandt, N., eds. (2011). *Experiments in Holism: Theory and Practice in Contemporary Anthropology*. John Wiley & Sons.

Overstreet, K. (2021). "Barn Cat Colonies in America's Dairyland Empty the Landscape." In *Feral Atlas: The More-Than-Human Anthropocene*, edited by A. L. Tsing, J. Deger, A. K. Saxena, & F. Zhou. Stanford University Press. https://feralatlas.supdigital.org/poster/barn-cat-colonies-in-americas-dairyland-empty-the-landscape

Paredes, A. (2021). "Chemical Cocktails Defy Pathogens and Regulatory Paradigms." In *Feral Atlas: The More-Than-Human Anthropocene*, edited by A. L. Tsing, J. Deger, A. K. Saxena, & F. Zhou. Stanford University Press. https://feralatlas.supdigital.org/poster/chemical-cocktails-defy-pathogens-and-regulatory-paradigms

Parsons, T. (1951). *The Social System*. Routledge & Kegan Paul.

Pegg, K. G., Coates, L. M., O'Neill, W. T., & Turner, D. W. (2019). "The Epidemiology of Fusarium Wilt of Banana." *Frontiers in Plant Science*, *10*, 1395. https://doi.org/10.3389/fpls.2019.01395

Perfecto, I. (2021). "Coffee Rust Spreads Together with Coffee Plantations." In *Feral Atlas: The More-Than-Human Anthropocene.*, edited by A. L. Tsing,

J. Deger, A. K. Saxena, & F. Zhou. Stanford University Press. https://feralatlas.supdigital.org/poster/coffee-rust-spreads-together-with-coffee-plantations

Pieterse, A. H. (1997). "*Eichhornia crassipes* (Martius) Solms." In *Plant Resources of South-East Asia: Vol. 11: Auxiliary Plants*, edited by I. F. Hanum & L. J. G. van der Maesen (pp. 118–121). Backhuys.

Piltch, E. M., Shin, S. S., Houser, R. F., & Griffin, T. (2020). "The Complexities of Selling Fruits and Vegetables in Remote Navajo Nation Retail Outlets: Perspectives from Owners and Managers of Small Stores." *Public Health Nutrition*, *23*(9), 1638–1646. https://doi.org/10.1017/S1368980019003720

Plumwood, V. (1993). *Feminism and the Mastery of Nature*. Routledge.

Porteous, J. D. (1986). "Intimate Sensing." *Area*, *18*, 250–251.

Price, R. (1990). *Alabi's World*. Johns Hopkins University Press.

———. (2002). *First-Time: The Historical Vision of an African American People,* 2nd ed. University of Chicago Press.

Pyne, S. (2021). "Fort McMurray Becomes a Portal to the Pyrocene." In *Feral Atlas: The More-Than-Human Anthropocene*, edited by A. L. Tsing, J. Deger, A. K. Saxena, & F. Zhou. Stanford University Press. https://feralatlas.supdigital.org/poster/fort-mcmurray-becomes-a-portal-to-the-pyrocene

Rackham, O. (2012). *Woodlands*. Collins.

Rae, L. (2021, October 21). "What Happens When You Remove a Dam? One Year Later, Life Returns to a Hudson River Tributary." *Riverkeeper*. https://www.riverkeeper.org/blogs/ecology/what-happens-when-you-remove-a-dam-life-returns/

Ramsbottom, J. (1937). "Dry Rot in Ships." *The Essex Naturalist*, *25*, 231–267.

Reilly, E. (2009). "Hence Mystical Cosmetic over Sunset Landfill." In *Styrofoam*. Roof Books.

Ren, J. (2021, February 25). "Wa and Fire, Old Village and New Village." *ThePaper.Cn*. https://www.thepaper.cn/newsDetail_forward_11459970

"A Resolution of the City of Flagstaff City Council Declaring a Housing Emergency for the City of Flagstaff, Prioritizing Affordable Housing Within City Operations to Create Safe, Decent, and Affordable Housing Opportunities for All Community Members." (2020). https://www.flagstaff.az.gov/DocumentCenter/View/71560/Housing-Emergency-Declaration---Resolution-2020-66

Richardson, D. M. (2021). "Biological Invasions of the Last 500 Years." In *Feral Atlas: The More-Than-Human Anthropocene*, edited by A. L. Tsing, J. Deger, A. K. Saxena, & F. Zhou. Stanford University Press. https://feralatlas.supdigital.org/index?text=richardson-invasion&ttype=essay&cd=true

Robin, V. (2019). "Dutch Elm Disease." In *Ridiculous Light*. Persea Books.

Rosaldo, R. (1980). *Ilongot Headhunting, 1883–1974: A Study in Society and History*. Stanford University Press.

Rose, D. B. (2012). "Multispecies Knots of Ethical Time." *Environmental Philosophy*, *9*(1), 127–140.

Rosenblum, E. B., James, T. Y., Zamudio, K. R., Poorten, T. J., Ilut, D., Rodriguez, D., Eastman, J. M., Richards-Hrdlicka, K., Joneson, S., Jenkinson, T. S., Longcore, J. E., Parra Olea, G., Toledo, L. F., Arellano, M. L., Medina, E. M., Restrepo, S., Flechas, S. V., Berger, L., Briggs, C. J., & Stajich, J. E. (2013). "Complex History of the Amphibian-Killing Chytrid Fungus Revealed with Genome Resequencing Data." *PNAS, 110*(23), 9385–9390. https://doi.org/10.1073/pnas.1300130110

Rose-Redwood, R., Barnd, N. B., Lucchesi, A. H., Dias, S., & Patrick, W. (2020). "Decolonizing the Map: Recentering Indigenous Mappings." *Cartographica, 55*(3), 151–162. https://doi.org/10.3138/cart.53.3.intro

Roy, B., Alexander, H. M., Davidson, J., Campbell, F. T., Burdon, J. J., Sniezko, R., & Braisier, C. (2021). "Increasing Forest Loss Worldwide from Invasive Pests Requires New Trade Regulations." In *Feral Atlas: The More-Than-Human Anthropocene*, edited by A. L. Tsing, J. Deger, A. K. Saxena, & F. Zhou. Stanford University Press. https://feralatlas.supdigital.org/poster/increasing-forest-loss-worldwide-from-invasive-pests-requires-new-trade-regulations

Sahlins, M. D. (1985). *Islands of History*. University of Chicago Press.

Savory, F., Leonard, G., & Richards, T. A. (2015). "The Role of Horizontal Gene Transfer in the Evolution of the Oomycetes." *PLOS Pathogens, 11*(5), e1004805. https://doi.org/10.1371/journal.ppat.1004805

Schmitz, H. (2021). "For a Hot and Humid Summer, I Traveled Through Georgia, Alabama and South Carolina." In *Feral Atlas: The More-Than-Human Anthropocene*, edited by A. L. Tsing, J. Deger, A. K. Saxena, & F. Zhou. Stanford University Press. https://feralatlas.supdigital.org/poster/for-a-hot-and-humid-summer-i-traveled-through-georgia-alabama-and-south

Schmitz, O. J., Buchkowski, R. W., Smith, J. R., Telthorst, M., & Rosenblatt, A. E. (2017). "Predator Community Composition Is Linked to Soil Carbon Retention Across a Human Land Use Gradient." *Ecology, 98*(5), 1256–1265. https://doi.org/10.1002/ecy.1794

Schmitz, O. J., Wilmers, C. C., Leroux, S. J., Doughty, C. E., Atwood, T. B., Galetti, M., Davies, A. B., & Goetz, S. J. (2018). "Animals and the Zoogeochemistry of the Carbon Cycle." *Science, 362*(6419), eaar3213. https://doi.org/10.1126/science.aar3213

Schranz, C. (2021). *Shifts in Mapping: Maps as a Tool of Knowledge*, Transcript Verlag. https://doi.org/10.1515/9783839460412

Schwinning, S., Belnap, J., Bowling, D., & Ehleringer, J. (2008). "Sensitivity of the Colorado Plateau to Change: Climate, Ecosystems, and Society." *Ecology and Society, 13*(2). https://doi.org/10.5751/ES-02412-130228

Scott, J. C. (1998). *Seeing Like a State: How Certain Schemes to Improve the Human Condition Have Failed*. Yale University Press.

Seeberg, J. (2021). "Bombarding Microbial Life with Antibiotics Creates an Explosion of Drug Resistance." In *Feral Atlas: The More-Than-Human Anthropocene*, edited by A. L. Tsing, J. Deger, A. K. Saxena, & F. Zhou. Stanford University Press. https://feralatlas.supdigital.org/poster/bombarding-microbial-life-with-antibiotics-creates-an-explosion-of-drug-resistance

Sen, A. (1982). *Poverty and Famines: An Essay on Entitlement and Deprivation*. Oxford University Press.

Singh, J., Bech-Andersen, J., Elborne, S. A., Singh, S., Walker, B., & Goldie, F. (1993). "The Search for Wild Dry Rot Fungus (*Serpula lacrymans*) in the Himalayas." *Mycologist*, 7(3), 124–130. https://doi.org/10.1016/S0269-915X(09)80072-4

Skelly, D., Arietta, A. Z. A., & Lambert, M. (2021). "Green Frogs Thrive in the Suburbs." In *Feral Atlas: The More-Than-Human Anthropocene*, edited by A. L. Tsing, J. Deger, A. K. Saxena, & F. Zhou. Stanford University Press. https://feralatlas.supdigital.org/poster/green-frogs-thrive-in-the-suburbs

Skipper, P., Pike, J., & May, N. T. (1996). *Ngurrara 1* [Acrylic on canvas].

Snelders, E., Huis in 't Veld, R. A. G., Rijs, A. J. M. M., Kema, G. H. J., Melchers, W. J. G., & Verweij, P. E. (2009). "Possible Environmental Origin of Resistance of *Aspergillus fumigatus* to Medical Triazoles." *Applied and Environmental Microbiology*, 75(12), 4053–4057. https://doi.org/10.1128/AEM.00231-09

Snow, N., & Witmer, G. (2021). "Introduced American Bullfrogs Are Invasive." In *Feral Atlas: The More-Than-Human Anthropocene*, edited by A. L. Tsing, J. Deger, A. K. Saxena, & F. Zhou. Stanford University Press. https://feralatlas.supdigital.org/poster/american-bullfrogs-are-disease-carrying-invasive-species

Soluri, J. (2021). *Banana Cultures: Agriculture, Consumption, and Environmental Change in Honduras and the United States*, rev. and updated ed. University of Texas Press.

Spahr, J. (2011). "Gentle Now, Don't Add to Heartache." In *Well Then There Now*. David R. Godine.

———. (2021). "Gentle Now, Don't Add to Heartache." In *Feral Atlas: The More-Than-Human Anthropocene*, edited by A. L. Tsing, J. Deger, A. K. Saxena, & F. Zhou. Stanford University Press. https://feralatlas.supdigital.org/poster/gentle-now-dont-add-to-heartache

Spivak, G. C. (1988). "Subaltern Studies: Deconstructing Historiography." In *Selected Subaltern Studies*, edited by R. Guha & G. C. Spivak (pp. 197–221). Oxford University Press.

Stadler, T., & Buteler, M. (2007). "Migration and Dispersal of *Anthonomus grandis* (Coleoptera: Curculionidae) in South America." *Revista de La Sociedad Entomologica Argentina*, 66(3–4), 205–217.

Steffen, W. (2021). "The Great Acceleration: The Collision of Human and Earth History." In *Feral Atlas: The More-Than-Human Anthropocene*, edited by A. L. Tsing, J. Deger, A. K. Saxena, & F. Zhou. Stanford University Press. https://feralatlas.supdigital.org/index?text=steffen-acceleration&ttype=essay&cd=true

Steffen, W., Crutzen, P. J., & McNeill, J. R. (2007). "The Anthropocene: Are Humans Now Overwhelming the Great Forces of Nature?" *AMBIO: A Journal of the Human Environment*, 36(8), 614–621. https://doi.org/10.1579/0044-7447(2007)36[614:TAAHNO]2.0.CO;2

Stein, S. (2021). "A Parasitic Plant Devastates Peasant Crops on Abandoned

Plantations in Southern Africa." In *Feral Atlas: The More-Than-Human Anthropocene*, edited by A. L. Tsing, J. Deger, A. K. Saxena, & F. Zhou. Stanford University Press. https://feralatlas.supdigital.org/poster/a-parasitic-plant-devastates-peasant-crops-on-abandoned-plantations-in

Stein, S., & Luna, J. (2021). "Toxic Sensorium: Agrochemicals in the African Anthropocene." *Environment and Society*, *12*(1), 87–107. https://doi.org/10.3167/ares.2021.120106

Stern, L. (2021). "Rats Destroy Flora, Fauna . . . and Figs." In *Feral Atlas: The More-Than-Human Anthropocene*, edited by A. L. Tsing, J. Deger, A. K. Saxena, & F. Zhou. Stanford University Press. https://feralatlas.supdigital.org/poster/rats-destroy-flora-fauna-and-figs

Stoetzer, B. (2021). "Pigs, Viruses, and Humans Co-Evolve in a Deadly Dance." In *Feral Atlas: The More-Than-Human Anthropocene*, edited by A. L. Tsing, J. Deger, A. K. Saxena, & F. Zhou. Stanford University Press. https://feralatlas.supdigital.org/poster/pigs-viruses-and-humans-co-evolve-in-a-deadly-dance

Stoler, A. L. (2001). "Tense and Tender Ties: The Politics of Comparison in North American History and (Post) Colonial Studies." *Journal of American History*, *88*(3), 829–865. https://doi.org/10.2307/2700385

———. (2009). *Along the Archival Grain: Epistemic Anxieties and Colonial Common Sense*. Princeton University Press.

Støvring Hovmøller, M. (2011). "The Increased Risk of Global Wheat Rust Pandemics: Putting Yellow Rust into Perspective." In *Fungal Diseases: An Emerging Threat to Human, Animal, and Plant Health: Workshop Summary* (pp. 252–263). National Academies Press. https://doi.org/10.17226/13147

Strathern, M. (1987). "Out of Context: The Persuasive Fictions of Anthropology." *Current Anthropology*, *28*(3), 251–281.

———. (2001). *The Gender of the Gift: Problems with Women and Problems with Society in Melanesia*. University of California Press.

Strivay, L., & Mougenot, C. (2021). "The Innocent European Rabbit Makes a Mess." In *Feral Atlas: The More-Than-Human Anthropocene*, edited by A. L. Tsing, J. Deger, A. K. Saxena, & F. Zhou. Stanford University Press. https://feralatlas.supdigital.org/poster/the-innocent-european-rabbit-makes-a-mess

Stukenbrock, E. H. (2016). "The Role of Hybridization in the Evolution and Emergence of New Fungal Plant Pathogens." *Phytopathology*, *106*(2), 104–112. https://doi.org/10.1094/PHYTO-08-15-0184-RVW

Sugarman, M. (2021). "The Garbage Can Is a Vessel Carrying Your Spent Consumables on the First Leg of a Global Voyage." In *Feral Atlas: The More-Than-Human Anthropocene*, edited by A. L. Tsing, J. Deger, A. K. Saxena, & F. Zhou. Stanford University Press. https://feralatlas.supdigital.org/poster/the-garbage-can-is-a-vessel-carrying-your-spent-consumables-on-the-first

Sultana, F. (2015). "Emotional Political Ecology." In *The International Handbook of Political Ecology* (pp. 633–645). Edward Elgar Publishing.

———. (2022a). "Critical Climate Justice." *The Geographical Journal, 188*, 118–124. https://doi.org/10.1111/geoj.12417

———. (2022b). "The Unbearable Heaviness of Climate Coloniality." *Political Geography, 99*, 102638. https://doi.org/10.1016/j.polgeo.2022.102638

Swanson, H. A. (2015). "Placing a Golden Spike at the Golden Spike: Railroads in the Making of the Anthropocene." In *Placing the Golden Spike: Landscapes of the Anthropocene*, edited by D. Hannah & S. Krajewski (pp. 102–111). INOVA (Institute of Visual Arts).

———. (2021). "An Explosion of Parasitic Lice Caused by Industrial Fish Farming Threatens Wild Salmon Populations." In *Feral Atlas: The More-Than-Human Anthropocene*, edited by A. L. Tsing, J. Deger, A. K. Saxena, & F. Zhou. Stanford University Press. https://feralatlas.supdigital.org/poster/an-explosion-of-parasitic-lice-caused-by-industrial-fish-farming-threatens-to-decimate-wild-salmon-populations

———. (2022). *Spawning Modern Fish: Transnational Comparison in the Making of Japanese Salmon*. University of Washington Press.

Syme, T. (2020). "Localizing Landscapes: A Call for Respectful Design in Indigenous Counter Mapping." *Information, Communication & Society*, 23(8), 1106–1122. https://doi.org/10.1080/1369118X.2019.1701695

Taijian, Q. (2015, January 16). "The Slow Life of Wengding Wa Primitive Tribe." *Chinese National Geography*. http://www.dili360.com/Article/p54b76e1a4725535/1.htm

TallBear, K. (2016, March 14). *Failed Settler Kinship, Truth and Reconciliation, and Science*. Courage and Social Justice in Our Time, University of Alberta. https://indigenousssts.com/failed-settler-kinship-truth-and-reconciliation-and-science/

Taussig, M. (1989). "History as Commodity: In Some Recent American (Anthropological) Literature." *Critique of Anthropology*, 9(1), 7–23. https://doi.org/10.1177/0308275X8900900102

Taylor, D. E. (2014). *Toxic Communities: Environmental Racism, Industrial Pollution, and Residential Mobility*. NYU Press.

TCMltd. (2010, May 21). *Japanese Knotweed*. https://www.youtube.com/watch?v=cobPduEnOVE

Téllez, T. R., López, E., Granado, G., Pérez, E., López, R., & Guzmán, J. (2008). "The Water Hyacinth, *Eichhornia crassipes*: An Invasive Plant in the Guadiana River Basin (Spain)." *Aquatic Invasions*, 3(1), 42–53. https://doi.org/10.3391/ai.2008.3.1.8

Tewksbury, J. J., Anderson, J. G. T., Bakker, J. D., Billo, T. J., Dunwiddie, P. W., Groom, M. J., Hampton, S. E., Herman, S. G., Levey, D. J., Machnicki, N. J., del Rio, C. M., Power, M. E., Rowell, K., Salomon, A. K., Stacey, L., Trombulak, S. C., & Wheeler, T. A. (2014). "Natural History's Place in Science and Society." *Bioscience*, 64(4), 300–310. https://doi.org/10.1093/biosci/biu032

Thomas-Blate, J. (2023, February 14). "Dam Removals Continue Across the U.S. in 2022." *American Rivers: River Stories*. https://www.americanrivers.org/2023/02/dam-removals-continue-across-the-u-s-in-2022/

Thompson, J. N. (2013). *Relentless Evolution*. University of Chicago Press.

Tinker, J. (1974). "Waterweeds: Flies in the Irrigation Ointment." *New Scientist*, 21(March), 747–749.

Trouillot, M.-R. (2011). *Silencing the Past: Power and the Production of History*. Beacon Press.

Tsai, Y.-L. (2016). "Agricultural Renaissance in Taiwan." *Router: A Journal of Cultural Studies*, 22, 23–74.

Tsai, Y.-L., Carbonell, I., Chevrier, J., & Tsing, A. L. (2016). "Golden Snail Opera: The More-Than-Human Performance of Friendly Farming on Taiwan's Lanyang Plain." *Cultural Anthropology*, 31(4), 520–544. https://doi.org/10.14506/ca31.4.04

Tsing, A. L. (1993). *In the Realm of the Diamond Queen: Marginality in an Out-of-the-Way Place*. Princeton University Press.

———. (2005). *Friction: An Ethnography of Global Connection*. Princeton University Press.

———. (2015). *The Mushroom at the End of the World: On the Possibility of Life in Capitalist Ruins*. Princeton University Press.

———. (2017). "A Threat to Holocene Resurgence Is a Threat to Livability." In *The Anthropology of Sustainability: Beyond Development and Progress*, edited by M. Brightman & J. Lewis (pp. 51–66). Palgrave Macmillan.

———. (2019). "The Political Economy of the Great Acceleration, or, How I Learned to Stop Worrying and Love the Bomb." In *Climate, Capitalism, and Communities*, edited by A. B. Stensrud & T. H. Eriksen (pp. 22–40). Pluto Press.

Tsing, A. L., Bubandt, N., Gan, E., & Swanson, H. A., eds. (2017). *Arts of Living on a Damaged Planet: Ghosts and Monsters of the Anthropocene*. University of Minnesota Press. https://doi.org/10.5749/j.ctt1qft070

Tsing, A. L., Deger, J., Saxena, A. K., & Zhou, F., eds. and curators. (2021). *Feral Atlas: The More-Than-Human Anthropocene*. Stanford University Press. feralatlas.org

Tsing, A. L., Mathews, A. S., & Bubandt, N. (2019). "Patchy Anthropocene: Landscape Structure, Multispecies History, and the Retooling of Anthropology. An Introduction to Supplement 20." *Current Anthropology*, 60(S20), S186–S197. https://doi.org/10.1086/703391

Tuck, E., & McKenzie, M. (2015). *Place in Research: Theory, Methodology, and Methods*. Routledge.

Turnbull, D., & Watson-Verran, H. (1989). *Maps Are Territories, Science Is an Atlas: A Portfolio of Exhibits*. Deakin University Press.

Tuttle, S., Moore, G., & Benally, J. (2008). *The Navajo Nation Quick Facts* (No. AZ1471). University of Arizona Cooperative Extension. https://extension.arizona.edu/sites/extension.arizona.edu/files/pubs/az1471.pdf

Uerta, S., & Flores, P. (2022). *Worlds of Grey and Green: Mineral Extraction as Ecological Practice*. University of California Press.

Ullstrup, A. J. (1972). "The Impacts of the Southern Corn Leaf Blight Epidemics of 1970–1971." *Annual Review of Phytopathology*, 10(1), 37–50. https://doi.org/10.1146/annurev.py.10.090172.000345

USDA. (2022, July 27). "Wheat Stripe Rust." *Agricultural Research Service*.

https://www.ars.usda.gov/midwest-area/stpaul/cereal-disease-lab/docs/cereal-rusts/wheat-stripe-rust/
Valencia, P. (2022, April 21). "109 Properties 'Impacted' by 20K-Acre Tunnel Fire, Officials Say; 30 Homes Burned." *AZ Family*. https://www.azfamily.com/2022/04/21/tunnel-fire-grows-more-than-20k-acres-more-crews-be-deployed-thursday/
van Dooren, T. (2019). *The Wake of Crows: Living and Dying in Shared Worlds*. Columbia University Press.
van Veen, J. (1962). *Dredge, Drain, Reclaim: The Art of a Nation*. Martinus Nijhoff.
Vann, M. G. (2021). "Colonial Sewers Led to More Rats." In *Feral Atlas: The More-Than-Human Anthropocene*, edited by A. L. Tsing, J. Deger, A. K. Saxena, & F. Zhou. Stanford University Press. https://feralatlas.supdigital.org/poster/colonial-sewers-led-to-more-rats
Vann, M. G., & Clarke, L. (2019). *The Great Hanoi Rat Hunt: Empire, Disease, and Modernity in French Colonial Vietnam*. Oxford University Press.
Vine, M. (2021). "Residents Inhale Settler-Colonial Histories in the Owens Valley, California." In *Feral Atlas: The More-Than-Human Anthropocene*, edited by A. L. Tsing, J. Deger, A. K. Saxena, & F. Zhou. Stanford University Press. https://feralatlas.supdigital.org/poster/residents-inhale-settler-colonial-histories-in-the-owens-valley-california
Viveiros de Castro, E. (2004). "Perspectival Anthropology and the Method of Controlled Equivocation." *Tipití: Journal of the Society for the Anthropology of Lowland South America*, 2, 3–22.
———. (2014). *Cannibal Metaphysics: For a Post-Structural Anthropology*, translated by P. Skafish. Univocal.
———. (2015). *The Relative Native: Essays on Indigenous Conceptual Worlds*. Hau Books.
Vodopivec, M., Kogovšek, T., & Malej, A. (2021). "Ocean Sprawl Causes Jellyfish Outbreaks." In *Feral Atlas: The More-Than-Human Anthropocene*, edited by A. L. Tsing, J. Deger, A. K. Saxena, & F. Zhou. Stanford University Press. https://feralatlas.supdigital.org/poster/ocean-sprawl-causes-jellyfish-outbreaks
Voeks, R., & Rashford, J., eds. (2013). *African Ethnobotany in the Americas*. Springer.
Wainwright, O. (2021, March 16). " 'Sometimes the Answer Is to Do Nothing': Unflashy French Duo Take Architecture's Top Prize." *The Guardian*. https://www.theguardian.com/artanddesign/2021/mar/16/lacaton-vassal-unflashy-french-architectures-pritzker-prize
Weiss, M. (2021). "Unexpected Threats to Trees Can Be Traced to Wood Pallets." In *Feral Atlas: The More-than-Human Anthropocene*, edited by A. L. Tsing, J. Deger, A. K. Saxena, & F. Zhou. Stanford University Press. https://feralatlas.supdigital.org/poster/unexpected-threats-to-trees-can-be-traced-to-wood-pallets
Weldon, C., du Preez, L. H., Hyatt, A. D., Muller, R., & Speare, R. (2004). "Origin of the Amphibian Chytrid Fungus." *Emerging Infectious Diseases*, *10*(12), 2100–2105. https://doi.org/10.3201/eid1012.030804

White, B. (1973). "Demand for Labor and Population Growth in Colonial Java." *Human Ecology*, *1*, 217–236.

Whittaker, K., & Vredenburg, V. (2011). "An Overview of Chytridiomycosis." *Amphibia Web*. http://amphibiaweb.org/chytrid/chytridiomycosis.html#references

Whyte, K. P. (2017). "Our Ancestors' Dystopia Now: Indigenous Conservation and the Anthropocene." In *The Routledge Companion to the Environmental Humanities*, edited by U. Heise, J. Christensen, & M. Niemann (pp. 206–215). Routledge.

Wilkinson, B. H. (2005). "Humans as Geologic Agents: A Deep-Time Perspective." *Geology*, *33*(3), 161–164.

Williams, A. E. (2005). "The Theory of Alternative Stable States in Shallow Lake Ecosystems." *Water Encyclopedia*. https://doi.org/10.1002/047147844X.sw24

———. (2006). "Water Hyacinth." In *Van Nostrand's Scientific Encyclopedia*, edited by G. D. Considine. John Wiley & Sons. https://doi.org/10.1002/0471743984.vse7463.pub2

Williams, C. (2020). *Intimate Sensing in Climate Research*. PhD dissertation. RMIT University.

Williams, E. (1944). *Capitalism and Slavery*. University of North Carolina Press.

Wilmers, C. C., & Schmitz, O. J. (2016). "Effects of Gray Wolf-Induced Trophic Cascades on Ecosystem Carbon Cycling." *Ecosphere*, *7*(10). https://doi.org/10.1002/ecs2.1501

Win, S., & Kumazaki, M. (1998). "The History of Taungya Plantation Forestry and Its Rise and Fall in the Tharrawaddy Forest Division of Myanmar (1869–1994)." *Japan Society of Forest Planning*, 4(1). https://www.burmalibrary.org/en/the-history-of-taungya-plantation-forestry-and-its-rise-and-fall-in-the-tharrawaddy-forest-division

Winther, R. G. (2020). *When Maps Become the World*. University of Chicago Press.

Wisconsin Department of Natural Resources. (n.d.). *Heterobasidion Root Disease (HRD)*. Wisconsin Department of Natural Resources. Retrieved April 8, 2023. https://dnr.wisconsin.gov/topic/foresthealth/annosumrootrot

Wolf, E. R. (1982). *Europe and the People Without History*. University of California Press.

Wolford, W. (2021). "The Plantationocene: A Lusotropical Contribution to the Theory." *Annals of the American Association of Geographers*, *111*(6), 1622–1639. https://doi.org/10.1080/24694452.2020.1850231

Worster, D. (1992). *Rivers of Empire: Water, Aridity, and the Growth of the American West*. Oxford University Press.

Wright, J. P., & Jones, C. G. (2006). "The Concept of Organisms as Ecosystem Engineers Ten Years On: Progress, Limitations, and Challenges." *BioScience*, *56*(3), 203. https://doi.org/10.1641/0006-3568(2006)056[0203:TCOOAE]2.0.CO;2

Wright, S. (2021). "I've Lost the Smell of Elm Dust." In *Feral Atlas: The More-*

Than-Human Anthropocene, edited by A. L. Tsing, J. Deger, A. K. Saxena, & F. Zhou. Stanford University Press. https://feralatlas.supdigital.org/poster/ive-lost-the-smell-of-elm-dust

Wynter, S. (1971). "Novel and History, Plot and Plantation." *Savacou*, *5* (June), 95–102.

———. (2015). "The Ceremony Found: Towards the Autopoetic Turn/Overturn, Its Autonomy of Human Agency and Extraterritoriality of (Self-)Cognition." In *Black Knowledge/Black Struggles: Essays in Critical Epistemology*, edited by J. R. Ambroise & S. Broek (pp. 184–252). Liverpool University Press.

Yang, L., & Zhao, F. (2021). "Land Use Determines Soil Antibiotic Concentrations." In *Feral Atlas: The More-Than-Human Anthropocene*, edited by A. L. Tsing, J. Deger, A. K. Saxena, & F. Zhou. Stanford University Press. https://feralatlas.supdigital.org/poster/land-use-determines-soil-antibiotic-concentrations

Young, M. (2021, November 19). "Celebrate the Wins for Rivers and Clean Water in the Infrastructure Investment and Jobs Act." *American Rivers: River Stories*. https://www.americanrivers.org/2021/11/celebrate-the-wins-for-rivers-and-clean-water-in-the-infrastructure-investment-and-jobs-act/

Zalasiewicz, J. (2017). "The Extraordinary Strata of the Anthropocene." In *Environmental Humanities: Voices from the Anthropocene*, edited by S. Opperman & S. Iovino. Rowman & Littlefield.

Zalasiewicz, J., Waters, C. N., Williams, M., Barnosky, A. D., Cearreta, A., Crutzen, P., Ellis, E., Ellis, M. A., Fairchild, I. J., Grinevald, J., Haff, P. K., Hajdas, I., Leinfelder, R., McNeill, J., Odada, E. O., Poirier, C., Richter, D., Steffen, W., Summerhayes, C., . . . Oreskes, N. (2015). "When Did the Anthropocene Begin? A Mid-Twentieth Century Boundary Level Is Stratigraphically Optimal." *Quaternary International*, *383*, 196–203. https://doi.org/10.1016/j.quaint.2014.11.045

Zee, J. (2021). "Mercury Fog Links China's Power Structure to the Deep Time of Coal Formation." In *Feral Atlas: The More-Than-Human Anthropocene*, edited by A. L. Tsing, J. Deger, A. K. Saxena, & F. Zhou. Stanford University Press. https://feralatlas.supdigital.org/poster/mercury-fog-links-chinas-power-structure-to-the-deep-time-of-coal-formation

Zee, J. C. (2022). *Continent in Dust: Experiments in a Chinese Weather System*. University of California Press.

Zhang, Y.-Y., Zhang, D.-Y., & Barrett, S. C. H. (2010). "Genetic Uniformity Characterizes the Invasive Spread of Water Hyacinth (*Eichhornia crassipes*), a Clonal Aquatic Plant." *Molecular Ecology*, *19*(9), 1774–1786. https://doi.org/10.1111/j.1365-294X.2010.04609.x

Zimmer, A. (2021). "In a Fossil Fuel Economy, Air Can Become Unbreathable." In *Feral Atlas: The More-Than-Human Anthropocene*, edited by A. L. Tsing, J. Deger, A. K. Saxena, & F. Zhou. Stanford University Press. https://feralatlas.supdigital.org/poster/in-a-fossil-fuel-economy-air-can-become-unbreathable

INDEX

Page numbers in *italics* indicate photographs.

IMAGE CREDITS

Figure 1: Image by Miyarrka Media. From Miyarrka Media, *Phone & Spear: A Yuṯa Anthropology*. Page xvii, 2019.
Figure 2. Photograph by Anna Tsing, 2014.
Figure 3. Photograph by Anna Tsing, 2013.
Figure 4. Photograph by Anna Tsing, 2018.
Figure 5. Photograph by Olga Kokcharova, 2018.
Figure 6. Photograph by Anna Tsing, 2017.
Figure 7. Photograph by Anna Tsing, 2017.
Figure 8. Photograph by Colin Hoag on behalf of the AURA Søby Brunkulslejer Research Group, 2018.
Figure 9. Photograph by Anna Tsing, 2014.
Figure 10. Detail of illustration depicting the stowage of British slave ship *Brookes,* published by the Plymouth Chapter of the Society for Effecting the Abolition of the Slave Trade, 1788.
Figure 11. Map by Lili Carr, 2021. Adapted from G. Korotaev et al., 2003.
Figure 12. Artwork by Liz Clarke. From M. Vann & L. Clarke, *The Great Hanoi Rat Hunt: Empire, Disease, and Modernity in French Colonial Vietnam,* 2018.
Figure 13. Diagram by Dave Brenner. Used courtesy of Ivette Perfecto.
Figure 14. Artwork by Victoria Baskin Coffey, 2021.
Figure 15. Photograph by Eduardo Grou, Universidade Estadual de Maringá, Paraná, Brazil. First published in F. Oda et al., 2019.
Figure 16. Still from video. Kishony Lab, at Harvard Medical School & Technion, Israel Institute of Technology, 2016.
Figure 17. Map adapted from Shipmap.org with source data from exactEarth and Clarksons ©Kiln.digital, 2008/2012.
Figure 18. Video still. Marcela Cely-Santos, Nicole Deger-Beauman, and Lili Carr, 2021.
Figure 19. Video still. *Greenhouse*, El Ejido, Spain. Armin Linke, 2013. Special thanks to Martina Pozzan.
Figure 20. Photograph by Pierre du Plessis, 2015.
Figure 21. Map by B. Lehrman. From "Visualizing Water Infrastructure with Sankey Maps," *Journal of Maps*, 2018.

Figure 22. Video still. *Natural Gas Extraction.* Qatar. Isabelle Carbonell and Duane Peterson, 2008.

Figure 23. Video still. *Corn Harvest.* Champaign County, Illinois. Bruce Rhoads and Trevor Birkenholtz, 2018. Special thanks to Armin Linke.

Figure 24. Map by Yukio Hayakawa. From *Yukio Hayakawa's Volcano Blog*, 2013.

Figure 25. Detail of *Capital.* Feifei Zhou, 2021.

Figure 26. *Invasion.* Feifei Zhou, incorporating original works by Nancy McDinny (*Story of Mayawagu*, 2013) and Andy Everson (*Heritage*, 2004), with permission, 2021.

Figure 27. *Yätj Garkman* (Evil Frog). Russell Ngadiyali Ashley, 2017. Acrylic on canvas.

Figure 28. Photograph by Michael Frohlich, courtesy of the Archaeology & Historic Preservation Division, State Historical Society of North Dakota, 2008.

Figure 29. *Empire.* Feifei Zhou, incorporating original work by Larry Botchway, *Contract*, in collaboration, 2021.

Figures 30–33. Details of *Empire.* Feifei Zhou, 2021.

Figure 34. *Alabama Fields.* Helene Schmitz. Kudzu Project, 2013. Photograph.

Figure 35. Image by EPP. Majavamm.jpg (cropped), 2011.

Figure 36. Image by Gerhard Elsner. Wurzelschwamm.jpg, 2003.

Figure 37. Image by Feifei Zhou, 2022. Drawing based on Asiegbu et al., 2005; Wisconsin Department of Natural Resources, n.d.; and Dambrauskaite, 2019.

Figure 38. Image by David B. Langston, University of Georgia, Bugwood.org, 2006.

Figure 39. Image from Depotter et al., 2016.

Figure 40. Image by Marek Argent, 2013.

Figure 41. *Yätj Garkman* (Evil Frog). Russell Ngadiyali Ashley, 2017. Acrylic on canvas.

Figure 42. *Cover the Earth.* Emmy Lingscheit, 2016. Lithograph with screen print.

Figure 43. Photograph by Andrew Mathews. Ancient chestnut stump, Pizzorna, Lucca, 2013.

Figure 44. Photograph by Rima Rantisi, 2019.

Figure 45. *Alabama Fields.* Helene Schmitz. Kudzu Project, 2013. Photograph.

Figure 46. Image by Kate Orff and Richard Misrach, Plate 32, *Petrochemical America*, 2012.

Figure 47. *Soil Map.* Frédérique Aït-Touati, Alexandra Arènes, and Axelle Grégoire. *Terra Forma: A Book of Speculative Maps*, 2022 (originally published as *Terra Forma: Manuel de Cartographies Potentielles*, 2019).

Figure 48. *WAKA WAKA GUDETAMANANGGAL.* Installation and photograph by Amy Lien and Enzo Camacho, 2016. Modified mannequin split into two halves and various materials.

Figure 49. *BRUTTO NETTO MANANANGGAL.* Installation and photograph by Amy Lien and Enzo Camacho, 2016. Modified mannequin split into two halves and various materials.

Figure 50. Detail of *Empire*. Feifei Zhou, incorporating original work by Larry Botchway, *Contract*, in collaboration, 2021.

Figure 51. *Contract*. Larry Botchway, 2020.

Figure 52. *Story of Mayawagu*. Nancy McDinny, 2013. Acrylic on linen.

Figure 53. Photograph by Zhiqiang Xiao, 2017.

Figure 54. Image by Miyarrka Media, 2019.

Figures 55–60. Video stills by Miyarrka Media, 2023.

Figures 61–68. Screenshots from *Feral Atlas: The More-than-Human Anthropocene*, digital project edited and curated by Tsing et al., 2021. CC-BY-ND-NC 4.0.

Figure 66 screenshot includes video still of *First Oil Well*, Jabal al-Dukhan, Bahrain. Armin Linke, 2014. Special thanks to Martina Pozzan.

Figures 67 and 68 also show partial images of a Center for Disease Control and Prevention map of countries in the world in which *Candida auris* has been detected, 2020.